Logical Universe

Logical Universe

A Layman's Reality

MICHAEL F. JONES

INERTIAL PRESS
Tacoma, WA

Logical Universe

All rights reserved.

Copyright © 2006 by Michael F. Jones
Original illustrations © 2006 by Michael F. Jones

Cover and interior design by Lightbourne, Inc. www.lightbourne.com

Cover and interior illustrations by Access Design, Inc.
www.nwbay.com

Space photos used in illustrations are courtesy of NASA/JPL-Caltech.

No part of this book may be reproduced or transmitted in any form
or by any means, electronic or mechanical, including photocopying, recording,
or by any information storage and retrieval system, without permission in writing from
the publisher, except by a reviewer who wishes to quote brief passages in connection
with a review written for inclusion in a magazine or newspaper.

Published by Inertial Press™
P.O. Box 45470 Tacoma, WA 98445

www.inertialpress.com

Library of Congress Control Number: 2005901687

Printed in the United States of America

ISBN-13: 978-0-9766247-7-6
ISBN-10: 0-9766247-7-X

CONTENTS

Preface • ix

CHAPTER ONE
THE ORIGIN AND COMPOSITION OF THE UNIVERSE

Time Is Only a Comparison · · · · · · · · · 3
Logic Is Simply a Reflection · · · · · · · · 7
Has the Universe Always Existed? · · · · · · 9
Before the Universe Existed · · · · · · · · 12
Empty Space Is Nothing · · · · · · · · · · 13
How the Universe Came into Existence · · · 16
An Introduction to Origo · · · · · · · · · · 17
Chapter One Highlights · · · · · · · · · · · 22

CHAPTER TWO
THE ORIGIN AND DISTRIBUTION OF MATTER

The First Production Run · · · · · · · · · 27
Energy Structures and Their Distribution · · · 29
The Inward and the Outward Forces · · · · · 30
What Holds It All Together? · · · · · · · · 31
Chapter Two Highlights · · · · · · · · · · · 34

CHAPTER THREE
GRAVITY AND THE WARPING OF SPACE-TIME

"An Attraction between Objects" · · · · · · · · 39
The Fabric of the Universe and the
    Gravitational Field · · · · · · · · · · · · · 41
The Energy Acting upon an Object Within
    and Without a Gravitational Field · · · · · · 43
Is Gravity the Result of a Pull or a Push? · · · · 46
Artificial Gravitational Fields and a
    Revolutionary New Propulsion System · · · · 47
Gravitons and Gravitational Waves · · · · · · · · 52
Inertia · 53
Chapter Three Highlights · · · · · · · · · · · · 56

CHAPTER FOUR
LIGHT AND ITS MOTION THROUGH THE UNIVERSE

What Is Light When It Is Not Being Observed? · 61
Why Light Travels at the Speed That It Does · · 62
The Generation of Light · · · · · · · · · · · · · 70
The Frequency of Light · · · · · · · · · · · · · 73
The Motion of Light · · · · · · · · · · · · · · · 74
Misinterpreted Experiments · · · · · · · · · · · 77
The Michelson-Morley Experiment · · · · · · · · 79
The Apparent Corpuscular Property of Light · · 81
The Diffraction-Box Experiment · · · · · · · · · 84
Edwin Hubble's Doppler Shifts (I) · · · · · · · 85
The Red-Shift Communication System · · · · · 89

How a Gravitational Field Affects Light · · · · 90
Edwin Hubble's Doppler Shifts (II) · · · · · · · 93
The Age of the Universe · · · · · · · · · · · · · 94
Why the Night Sky is Dark · · · · · · · · · · · 95
Chapter Four Highlights · · · · · · · · · · · · · 97

CHAPTER FIVE
BLACK HOLES AND MATTER AT THE SPEED OF LIGHT

What a Black Hole Is · · · · · · · · · · · · · · · 101
Escape Velocity · · · · · · · · · · · · · · · · · · 101
The Ever-Changing Event Horizon · · · · · · · 103
Matter at the Speed of Light · · · · · · · · · · 105
The Light-Speed Limit · · · · · · · · · · · · · · 106
Matter above the Speed of Light · · · · · · · · 106
Chapter Five Highlights · · · · · · · · · · · · · 108

CHAPTER SIX
AN ATOMIC EXPLANATION

The Current Atomic Model · · · · · · · · · · · 111
Something Here Is Not Quite Right · · · · · · 111
The Optional Laws of Physics · · · · · · · · · 112
The Nucleus' Gravitational Field · · · · · · · · 114
Transient Energy and the Electrosphere · · · · 115
No Electrons Inside · · · · · · · · · · · · · · · 119
Origo's Motion through Matter · · · · · · · · · 122
Chapter Six Highlights · · · · · · · · · · · · · · 124

CHAPTER SEVEN
THE MAGNIFICENT MAGNETIC FIELD

The Energy Within a Magnetic Field · · · · · · 129
How Magnetic Fields Are Generated · · · · · · 131
Repulsion between Two North Poles · · · · · · 135
Repulsion between Two South Poles · · · · · · 137
Attraction between a North and a South Pole · · 138
Magnetic Monopoles · · · · · · · · · · · · · 140
Chapter Seven Highlights · · · · · · · · · · · 142

Conclusion • **143**

Acknowledgements • **145**

Predico • **151**

Gravitational Challenge • **155**

Glossary • **157**

Index • **167**

About the Author • **173**

Contact the Author • **177**

Order form • **178**

PREFACE

When I was a young boy, my Great Aunt Dorothy gave my brother Earl and me a set of science encyclopedias that were so interesting I couldn't stop reading them; they became prime entertainment for me. I suppose that is when my fascination with the universe began. I can remember, within a few years of receiving the encyclopedias, walking to my neighborhood library in search of two pieces of information that I felt I should know. Does the sun rotate, and why does gravity occur? I wasn't quite sure why I didn't know these things; perhaps it was because my books were out of date. Maybe I had not been paying attention in class. Possibly these subjects just hadn't been covered yet.

With full confidence that finding the answers to these questions was as simple as finding where the right book was, I strolled into the Moore Branch Library (where I have lost several homemade bicycles for failing to lock them up outside). After hours of fruitless searching, I left the library with the shocking realization that scientists did not understand the universe. I was able to find very little information about the sun and only superficial information about gravity; descriptions of what it is but not why.

Since that day I have found that the sun does indeed rotate (my library just didn't seem to have this information on hand), and the cause of gravity is still unknown to science. All popular theories about the universe are incomplete and/or illogical. The two most prevalent are Relativity and Quantum Theory.

In 1915 Albert Einstein gave us his General Theory of Relativity, which has taken us a long way on the road to understanding our universe. It has, among other things, revealed that matter and energy are the same thing, and it has described the gravitational field mathematically. The General Theory of Relativity, however, is incomplete. Like many previous theories and laws it does a very good job of explaining that things happen and explaining a bit about what is happening, but it does not explain why things such as gravitational fields occur. Isaac Newton also described gravitational phenomena, among other things such as inertia, but, like Einstein, he failed to explain the extremely important "why."

Quantum Theory attempts to explain physical reality in terms of the probability that events will happen or not, with no one being able to have certainty about these events. It is an exercise in statistical mathematics and these statistics seem to do well in helping us understand certain events, but when the statistics are presented as actual reality, scientific understanding is cast into disarray. The result is illogical beliefs like the idea that a single photon can go through two separate slits at the same time and that light can be waves or particles but not both. It also allows for illogical oddities such as Schrödinger's cat, which is said to be both alive and dead at the same time until it is observed, at which time the cat will become one or the other.

However, quantum theory, like Relativity, cannot tell us why things happen. This theory, which has been evolving for about a century, paints a picture of the universe that is far from reality. Although the statistical mathematics of quantum theory has made possible many scientific advances, it is incomplete, illogical, and fails to explain one of the most observable forces of the universe, gravity.

Gravity, inertia, magnetic fields, the propagation of light, and other phenomena all have the same question still demanding an answer: why?

There is reality at the base of everything that has yet to be understood, and people everywhere are yearning to understand it. Modern science, however, has alienated many people by presenting the universe as such an extremely complex thing that one cannot hope to understand it. Many people are so alienated and intimidated that they will not even try to understand. *Logical Universe* offers a common sense picture of the universe that everyone can understand.

The *Logical Universe* project began one evening in 2001 when, after years of personal studies and observations, I put a few lines down on paper in an attempt to develop a reasonable explanation for gravity that I could live with. I was not satisfied with the feeble explanations that modern science had to offer, which were little more than observations of what happens, with no explanation why.

The realization that I am not simply in the universe, but part of the universe, compels me to understand it. It is untenable for my consciousness to simply exist and then dissipate, without ever having had this understanding.

The first thing I did was to write down what was known

about gravity, omitted what was commonly assumed, and considered what possibilities may exist that could be the cause of gravity. It soon became clear that gravity could not be explained by itself. Light, matter, inertia, magnetism, and all other phenomena are indicators of how the universe works and must necessarily be considered in order to understand gravity. Therefore, my explanations for these things developed along with my explanation for gravity.

By the time I was content with my explanation of why gravity occurs, I realized that others may want to consider it and that there was enough information to organize into a book.

Modern science is much like the six blind men of India who all touched a different part of an elephant and who describe the elephant by the characteristics of the parts that they find. The blind man who finds the elephant's side thinks the elephant is like a wall, the one who finds the tusk thinks the elephant is like a spear, the one who finds the trunk thinks the elephant is like a snake, the one who finds the knee thinks the elephant is like a tree, the one who finds the ear thinks the elephant is like a sail, and the one who finds the tail thinks the elephant is like a rope. The reality is that while the elephant has parts that possess the characteristics described by the blind men, the entire elephant cannot be realized by examining any individual part.

Scientists, like the blind men, are all looking at different parts of the universe and trying to describe it by the characteristics of these parts. The main reason that our universe is not yet understood seems to be that no one is stepping back far enough to get a good look at the elephant as a whole.

Scientific experiments and observations can be carried out with the utmost precision, but no matter how accurate the experiment or observation, there is always one component that is subject to error—the human mind. The interpretation of experimental results or observations is the wild card.

All science that rests on a foundation of misinterpretation is inaccurate. Because modern science is largely based upon the inaccurate interpretation of previous experiments and observations, much of what we hear from the scientific community today is also inaccurate.

Analogous to the faulty construction of our scientific understanding is the faulty construction of a massive structure. When we hear the news of a massive building or bridge collapsing under its own weight, we are shocked. After the shock wears off, we wonder how such a thing could possibly happen. We are in disbelief that it could happen at all. Numerous experts were involved in the design and construction of the structure, just as there are numerous experts involved in the construction of our scientific understanding. We refuse to believe that all those engineers, architects, professional builders, and inspectors involved in the project did not discover the fatal flaw.

Modern science is much like that massive structure. It has fatal flaws that will be acknowledged only when the walls come crashing down. When people finally realize how flawed modern science is, many will stand in disbelief, just as if a great structure had collapsed.

Everybody has a personal world view. The world and, indeed, the universe around us are perceived differently by each one of us. Although there may be many similarities between these world views, we all exist in our own world; our individual realities are in our own minds. For example, most people may agree that the sky is blue, but each person's perception of blue may be slightly different.

When something comes along that threatens our world view, we tend to resist it; for many such a threat can be frightening. Because of this tendency, it is possible that, at least subconsciously, scientists tend to observe and explain things in such a way as to protect their personal world views.

Few things are as satisfying as understanding our universe. *Logical Universe* makes this easy by presenting simple and logical explanations for many aspects of our universe, including gravity, magnetic fields, inertia, and the reason that light travels at the speed that it does. *Logical Universe* also presents a bold, new, logical explanation for blue- and red-shifted light from distant galaxies, which shows that this phenomenon indicates the amount of matter in a given volume rather than the expansion of the universe.

Logical Universe is a book for everyone who wants to understand their universe.

Read and enjoy!

Michael F. Jones

CHAPTER ONE

THE ORIGIN AND COMPOSITION OF THE UNIVERSE

In this chapter

- We will examine the concepts of time and logic
- Consider how the universe began
- Explore the notion of empty space
- Reason logically how the universe came into existence
- Be introduced to an energy called origo

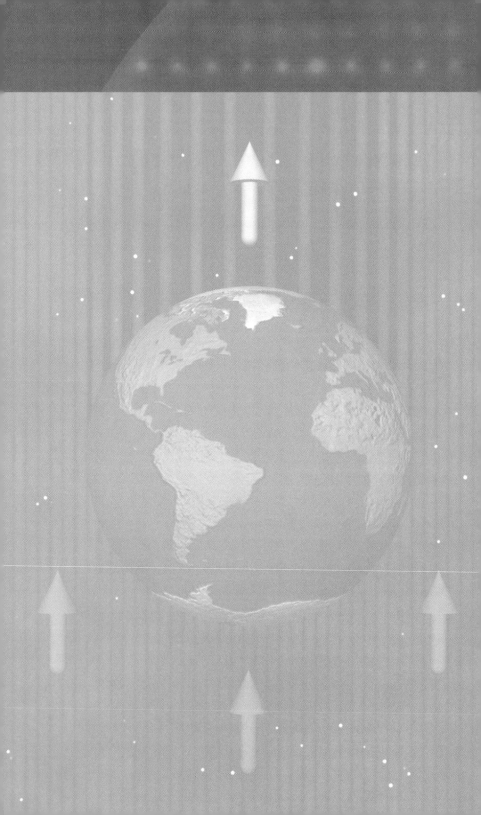

Time Is Only a Comparison

Time seems to be a mysterious thing that moves in a forward direction. We are thrust into the future constantly, whether we welcome it or not, and we attribute this to time as if it were some tangible thing pushing us along.

Some people speak about time's curious property that allows us to divide it into infinitesimally small periods. The fact is, however, that we do not divide time.

Time is the name used to describe the system of measuring either how long events take to occur or the length of a period between particular events. The length of these periods will not change, no matter what we call them.

The names that we give to certain periods, such as seconds, minutes, hours, etc., enable people to communicate with one another about how long the intervals between events are and how long events take to happen. If we did not use these universally understood terms, one person could not say to another, "Meet me in an hour," and expect to meet the other person at the expected time. Under these conditions, the world would be a very confused place.

Some people say that time is unidirectional, always flowing relentlessly toward the future. The fact is, time does not flow in any direction. We typically have a distorted view of what time really is. It is often thought of and portrayed as some entity that exists independent of everything else in the universe.

Our perception of time is a direct result of the logical way that our universe works. We see time as flowing forward because that is how events seem to occur. When a ball falls through the air and strikes the ground, we see the ball fall through the air and then strike the ground. For us, this is the way it is supposed to happen.

We know from previous experience that the ball cannot strike the ground before falling through the air. In this universe, when events happen, they happen in the logical manner that we have come to expect from previous experience.

Events always happen in an order which, from our perspective, seems like a forward motion, as each event always precedes the next logical event in a sequence. The entire event or sequence of events happens within a certain time period. In our minds we couple that time period, or what we perceive to be that time period, with what we perceive as forward-moving events, giving us the illusion that time is some independent entity in motion. This same type of illusion gives us the feeling that "time is standing still" or that "time is flying by."

Occasionally, we perceive the period in which events happen as being shorter or longer than it actually is. When time seems to stand still, we perceive an event or sequence of events happening within a longer period than it actually does. Likewise, when time seems to fly by, we perceive an event or sequence of events happening within a shorter period than it actually does.

When we refer to the measurement of time, what are we actually saying? If someone were to say, "I will use my stopwatch to record how much time it takes for this event to happen," what he would actually mean is, "I will use my stopwatch to record how long this event takes to happen, in comparison to the period my stopwatch is calibrated to measure," which is the Earth rotating on its axis. In this example, which is representative of all "time measurements," some independent entity called time is not measured. What transpires is simply a comparison of the periods of two separate events: the event being observed and the rotation of the Earth. Time is nothing more than the measurement and comparison of the periods of separate events.

The period in which events are happening we call the present. From our perspective, the present is the here and now; it is a fleeting moment that was in the future and will soon be in the past. In the present and only in the present are events actually occurring.

Events do not occur in the future, and events do not occur in the past. Events that have already occurred did so within a certain period; when the event is finished, the period in which it occurred is also finished. This period is not something that continues to exist; it was merely the reality of how long the event took to occur. This period was a certain length, and if we compare it with another event or sequence of events, and record the results of this comparison, we would have knowledge of the length of the period. If we then proceed to continually compare the period from when the event occurred to the present and record it, we would have knowledge of when the event occurred relative to any event or period of which we also have knowledge. This knowledge gives us the illusion that the event and the physical objects associated with it somehow exist in the past; this is only an illusion.

Events occur only in the present; therefore, physical objects and reality exist only within the present. The past is nothing more than events that occurred before the events that are occurring right now in the present.

The future is something we see as a time yet to come. The future, however, is not a "time" at all.

We know from previous experience that events happening in the present will cause other events to happen. We are accustomed to mentally connecting these anticipated events with the periods within which they will occur. We do this out of habit and necessity, because if we did not, we would not survive for long, either as individuals or as a species. In order for people to survive, we must be able to predict the future. If we could not predict that walking in front of a fast-moving vehicle will result in serious injury or death, there would be no reason for us to refrain from doing so. In this world of fast-moving vehicles, we would be extremely lucky to survive for even one day.

Every action that we consciously perform is an exercise in predicting the future. The more variables that we are able to rapidly factor in, the better we become at predicting the immediate future, which increases our chances for short-term survival. In addition, the longer the sequence of events that we are able to accurately predict, the greater are our chances for long-term survival.

When events happen in the present, their effects in the future are inevitable.

There will be a certain period between particular present and future events. We can predict the length of this period, and therefore we imagine it as being a certain fixed length of time in the future, as if time is something that somehow extends from the present into the future. Time, as was stated earlier, is nothing more than the measurement and comparison of the periods of separate events. We can compare the periods within which events happen and the periods between events, but events occur only in the present. If we accept these statements as correct, it becomes clear that time is not some tangible thing that extends into the future; we only imagine that it is.

The future is nothing more than the inevitable events that will occur after the events that are occurring right now, in the present.

We all have probably read books or seen movies that deal with the subject of time travel. The classic idea of time travel involves people and/or other physical objects somehow "traveling through time" to a period in which the events taking place either had already occurred or had not yet occurred before the "time travel" took place. This idea assumes that time is some tangible thing that can be navigated within.

The idea that physical objects can travel into the past or future is illogical. It is illogical because there are no physical past or future realities into which to travel.

Logic Is Simply a Reflection

Human reasoning is the reflection of the nature of the universe. We tend to think logically because that is what we learn from observing the universe around us.

We know that a ball that is dropped from a height must fall through the air before it strikes the ground. We know this because we have seen it happen many times, and not only have we seen it many times, but we have seen it happen every time that a ball has fallen from a height. We have never observed a deviation from the sequence of events that we have come to expect, once the ball is dropped. The ball always falls through the air before it strikes the ground.

These statements may seem somewhat unnecessary because of their obviousness, but consider if the universe did not operate the way that it does, and the ball must instead strike the ground before it falls through the air. In a universe such as that, the statement that "the ball must fall through the air before it strikes the ground" would be illogical. It would be illogical because the nature of that universe would require the ball to strike the ground before falling through the air. Everyone in that universe would already know from experience that things do not happen in that order. This statement would be akin to the statement that "the ball must strike the ground before it falls through the air," taken in the context of the nature of our universe. For us, this is an illogical statement, but it is illogical simply because the nature of our universe dictates that the event preceding the ball striking the ground is the ball falling through the air.

These examples are meant to show that logic is not a set of rules of its own but is simply reasoning that coincides with the nature of the universe.

Because logic is reasoning that coincides with the nature of the universe, everything that exists and happens in the universe must be logical. Conversely, illogical things or events cannot exist in the universe.

Has the Universe Always Existed?

Why is the universe here and where did it come from are questions that the human race has asked for millennia. A multitude of answers to these questions has arisen over the years, all of which are illogical or incomplete. Even the most widely accepted theories leave something out, such as how the universe began and why.

From almost all points of view, the origin of the universe seems to be a very illogical thing; a good place to start on this subject is with a question: "Is it logical for the universe to have always been in existence?"

In an attempt to answer this question, let us first consider two scenarios regarding the origin of the universe. The first one involves a universe that began at a first event.

A first event is a single event that precedes all others.

If there was a first event, it would have set in motion a chain of events, such that if one could find a way to follow them, any event at any point in this chain could be traced back to the first event. This scenario allows events to happen in a linear fashion; they start at a point, and, in a manner of speaking, move away from the starting point. This universe would have had a beginning, and the first event would have been the beginning.

The second scenario involves a universe that did not begin at a first event. A universe such as this would not have a beginning; it would have always been in existence.

In a universe with no first event, events cannot happen in a linear fashion. Events of the future cannot simply disappear into the future with no connection to the past.

If there was no first event, then all events do have a preceding event, because the causes of all events are other events that preceded them.

The only configuration for a universe that has always existed is a closed loop. In a closed loop, all events are the cause of all other events. This means that the past and the future are one and the same. In this configuration, events that have not yet happened are the cause of all past events.

Even though we can reason how both of the above scenarios must operate, this does not mean that it is logical for them both to exist. The idea of a universe that began at a first event is a logically correct idea. Although the question of how and why the first event occurred has yet to be answered, there is nothing illogical about the sequence of events happening from the first event "forward."

The idea of a universe in which all events form a closed loop is a good example of the human mind's capacity to imagine illogical things. It is easy to imagine the idea of a scenario where everything that happens culminates in a single event or in a condition that causes all events to happen the same as identical events that have previously happened. The odds against such a scenario happening are so great that they may be unimaginable, and in addition, this scenario would be in violation of the second law of thermodynamics, which states that the disorder of an isolated system can never decrease.

For a closed loop universe, which would be an isolated system, to exist, the disorder of the system would have to fluctuate up and down. It would be more disorderly at some points and less disorderly at others. However, if in violation of the second law of thermodynamics such a scenario did develop where every event of the universe repeated itself in a way identical in every detail to the previous cycle, then events would still not form a closed loop. Although events in every cycle may be identical in almost every way, one happened before or after the other in the sequence. Therefore, because there is a time period between both events, each event is separate from its counterparts in the previous and future cycles.

For a closed loop to form, the same event must happen repeatedly.

Therefore, the idea of a self-perpetuating universe within which events form a closed loop is illogical. Because it is illogical for the universe to exist as a closed loop, the only remaining scenario is the logical one, in which the universe came into existence at the first event, from which all other events originate.

In our logical universe, the causes of all events, with the exception of the first event, are other events that precede them. Therefore, we can answer our earlier question. No, it is not logical for the universe to have always been in existence; it must have had a beginning.

Before the Universe Existed

We reasoned in the previous section that logic dictates that the universe must have had a beginning.

If the universe had a beginning, what was here before it came into existence? It is clear that if the universe had a beginning, there could have been nothing before it came into existence.

According to some, questions about conditions before the universe existed are senseless. This reasoning probably has to do with the logic that if the universe did not yet exist, then time did not yet exist; therefore, there was no "before."

The reasoning that there was no "before" makes sense, but if one is inclined to accept the idea that there is nothing to know about the pre-universe simply because nothing existed, logic will not support this stand.

There are several conclusions that one can arrive at by logical reasoning on the subject of the pre-universe, and if one is inclined to ponder the difficult idea of nothing, he or she may be pleasantly surprised to find more there than meets the eye.

The idea of nothing is difficult to write about because just to mention nothing is to imply a thing, which is exactly opposite of the idea that we are considering. Therefore, as we forge ahead, please keep this difficulty in mind. Nothing is just that, nothing: no matter, no energy, and no time.

Although the pre-universe may have been nothing, further reasoning will reveal the single most important fact about nothing, which allows for the existence of the universe. If there was absolutely nothing, then there could have been no governing laws, and therefore, there was no logic.

Empty Space Is Nothing

Could it be that what we think of as nothing and what we think of as empty space are, in fact, the same?

    When we imagine nothing, we expect:
    1. No mass.
    2. No energy.
    3. No shape.
    4. No time.

    When we imagine empty space, we expect:
    1. No mass.
    2. No energy.
    3. No shape.
    4. Infinite time.

If empty space were indeed empty, matter would not exist, energy would not exist, it would have no boundaries and would therefore be infinite. If an object were in empty space, we could measure the object's height, width, and depth, but we could not measure empty space, because it has no boundaries; there is no value to place on its length in any direction. Therefore, empty space would have no shape, because it is infinite.

Our perception that empty space must have infinite time stems from the thought of a clock ticking away as no events take place for eternity.

As we reasoned in the previous section about time, "time is nothing more than the measurement and comparison of the periods of separate events." If empty space contains no matter or energy, then time does not exist in empty space, because there are no events taking place that can be measured.

When we imagine a clock measuring time for empty space, what we are imagining is illogical. The clock that we imagine does not "measure time." It only makes a comparison between eventless empty space and the period that it takes an Earth that does not exist in this empty space to rotate on its non-existent axis.

For us to imagine empty space as having infinite time is illogical. As empty space would have no events to measure, it would have no time.

Now that we have resolved the time issue of empty space, the only obstacle in identifying it as "nothing" is its apparent property of having volume.

Does volume itself constitute a thing? If volume were a thing, then it should have one or more of the following properties:

1. It should have mass.
2. It should have energy.
3. It should have shape.

Volume itself has none of these properties; therefore, we cannot logically consider it a thing. We can only consider volume itself to be a concept.

We have reasoned that empty space has no mass, energy, shape, or time. By this reasoning, it fits the description of "nothing." The only possible discrepancy was our idea that it has volume, which we have reasoned does not constitute a thing, but is only a concept. Therefore, we can conclude that our concept of empty space matches exactly with our concept of nothing, once we realize that our ideas of time and volume being things in themselves are illogical.

When we think about the space between stars and galaxies, we typically consider it to be empty space. Space in our universe is not the same as the "empty space" discussed in the preceding paragraphs. In our universe there is no empty space, and in the following paragraphs we will discover why.

Anywhere that the quantity of matter per unit of volume is extremely low, we perceive there to be empty space.

The idea of empty space allows us to imagine that if we could remove everything from a specific volume, there would remain only empty space. This concept is correct; however, in our universe, we cannot actually accomplish this.

Consider a rock that is in deep space, many light years from any galaxies, a place where even a single atom might be a rare find. To many, this would be empty space, but it is not; it is only a place where there is little or no matter. There is light from galaxies in all directions that are reflecting off the rock. If the space were empty, there would be no light reaching the rock because the light would not be there. If we were to put the rock inside of a box that light could not penetrate, no light would reach the rock, but the space inside the box surrounding the rock is still not empty. There is gravity between the rock and the distant galaxies.

Although science cannot yet tell us what causes gravity, the fact is, it exists and permeates the entire universe. Therefore, before we can claim that there is empty space anywhere in the universe, we must find or make a place where even gravity does not exist.

The fact that gravity influences objects through space is direct evidence that space is not empty.

How the Universe Came into Existence

As we reasoned earlier, it is illogical for the universe to have always existed; it must have had a beginning.

If the universe had a beginning, then before the universe began there must have been nothing. It is logical that if there was nothing, there could not have possibly been any means with which to bring the universe into existence.

Logic tells us that if there were no means with which to bring the universe into existence, then it would be impossible for this to happen. This seems to be a dilemma, because we know that the universe does exist; therefore, something must have happened that brought it into existence. To go from nothing to something, some change would have to take place.

Do not despair! There is a logical way out of this dilemma. As we reasoned in the previous section, "Before the Universe Existed," if there was absolutely nothing, then there could have been no governing laws, and therefore, there was no logic. This is the gateway to the universe.

If there was nothing and therefore no governing law, then obviously, at this point, logic that we are familiar with did not apply. This means that it was possible for the universe to suddenly and spontaneously come into existence.

In the absence of governing laws, "nothing" was very unstable. This very unstable "nothing" changed because it could. There were no governing laws to make it change, but there were also no governing laws to make it remain unchanged; therefore, a change was possible.

"Nothing" suddenly and spontaneously changed to something, and the universe came into existence. This was the first event.

Let us look through the eyes of an observer who, for the sake of reasoning, can observe without having an effect on anything (or nothing). The observer sees that there is nothing, and she watches for countless millennia (relative only to the observer and her logical viewpoint) with no change happening. Suddenly, a change takes place and the universe comes into existence. The observer asks," Why did nothing remain unchanged for so long and then suddenly the universe came into existence?" The answer to this question would be that time passed only for the observer. Before this first event there was nothing; therefore, events were not occurring and time did not exist.

Although no time actually passed, had there been an isolated observer recording her own personal time until the universe came into existence, she would have recorded an infinite amount of time. If the observer began recording her own personal time when the universe came into existence but observed and recorded it backward, she would be doing it for eternity.

At the instant the universe came into existence, "nothing" ceased.

An Introduction to Origo

When the universe began, what was it that came into existence? This was the first event, and there were no previous governing laws to dictate what the universe should be like; therefore, it could have been any of all possible states. There also were no governing laws to predetermine any details of the new universe; therefore, unless details were an integral part of the possible states that the universe could have become, the universe would not have come into existence with any details.

Let us consider what states were possible for the new universe. The empty space that we identified as "nothing" before the universe existed was infinite; therefore, in order for the universe to come into existence at some size less than infinite, there needed to have been some limiting factor.

Because, pre-universe, there was nothing and therefore no limiting factor, the infiniteness of "nothing" was the default template for the universe. Therefore, the first condition for any and all possible states of the universe was that the universe must be infinite.

Pre-universe, there were no governing laws. Only after the universe came into existence could governing laws begin to exist.

Because there were no governing laws before the universe came into existence, there were no factors to determine any details; therefore, we can expect that the new universe had no details. After the universe formed and governing laws came into existence, details began to develop.

When the universe came into existence it was infinite and without details, which means that it would have been homogeneous. So far, we have an infinite, homogeneous universe that suddenly and spontaneously came into existence, but we don't know what came into existence.

Pre-universe there was nothing. Therefore, we can reason that when the universe came into existence it was something.

Because, pre-universe, there was nothing but infinite empty space, there was nothing to determine any details of the new universe, and therefore no reason for this something to be any certain type of substance; therefore, it was no certain type of substance. Even if it was some generic type of substance or energy, there must have been something to determine what it would be. However, the only determining factors were those default factors that existed due to the infiniteness of nothing and the lack of factors other than the infiniteness of nothing: namely, infiniteness and homogeneity.

There is only one thing that fits within the parameters of these determining factors: motion—infinite, homogeneous motion.

When motion came into existence it was infinite, and because there were no limiting factors, this motion came into existence with infinite velocity.

Motion itself is pure energy.

Motion doesn't really seem like something that can be a thing all by itself. We usually associate motion with matter or other forms of energy.

Motion by itself is a strange concept; one may ask "the motion of what?" or "what is in motion?" or "how can you have motion if there is nothing in motion?"

Thanks to Albert Einstein, we know that matter is energy. If we put a quantity of matter in motion, it gains a quantity of energy equal to that used to put it in motion; we call this kinetic energy.

In order to eliminate the possibility of using certain words for more than one meaning while describing this infinite motion, or energy, we will refer to it as origo.

The word origo is of Latin origin and is equivalent to the English word "source."

When the universe came into existence, what came into

existence was origo. Origo was infinite, and empty space ceased.

The universe is not filled with origo; the universe is origo.

When origo came into existence it was infinite motion. Infinite motion is an exciting concept; it acts in all directions at an infinite speed. With infinite speed, any distance, even an infinite distance, can be crossed instantly. Infinite motion acts in infinite directions. This means that origo travels in all directions at every point in the universe.

When the universe came into existence, the quantity of motion per unit of volume was infinite; origo was infinite speed acting in infinite directions and occupying an infinite volume; the motion could become no greater in any way. Therefore, its density was as great as it could possibly be.

Imagine a homogeneous field of motion as wide as the universe and as long, acting in a single direction; this is origo in one direction. Now imagine this same energy field acting in every direction; this is the motion of origo. Bear in mind that the quantity of different directions is infinite.

Origo, which came into existence as infinite motion, was acting in infinite directions everywhere in the universe. There is an apparent problem with this picture. If origo were motion at its greatest possible density, acting in every direction, it would not be able to move at all, because motion in every direction would be countered by motion in the opposite direction.

If origo, which is infinite motion, cannot move, then how can we call it motion and how can it move at an infinite speed? This dilemma apparently threatens the foundations of the origo theory. However, as we proceed, it will become clear that it is vital to the formation of the universe as we know it.

In order for origo to occupy an infinite volume, be infinite velocity, and be homogeneous, it must be completely fluid.

Being completely fluid means that it does not consist of particles or waves, but consists of motion (pure energy), which cannot be reduced to a smallest quantity. It consists of an infinite quantity of infinitesimal quantities of energy.

> **DID YOU KNOW?**
>
> **Planet Earth is thought to be 4.5 billion years old!**

CHAPTER ONE HIGHLIGHTS

1. Time is nothing more than the measurement and comparison of the periods of separate events.

2. Events occur only in the present.

3. Logic is reasoning that coincides with the nature of the universe.

4. The universe had a beginning with the first event.

5. Pre-universe, there were no governing laws; therefore, our logic did not apply.

6. Empty space and nothing are one and the same.

7. Volume is only a concept.

8. There is no empty space in our universe.

9. The universe suddenly and spontaneously came into existence.

10. The universe is infinite in size.

11. Governing laws and logic exist only because the universe exists.

12. The universe consists of infinite, homogeneous motion called origo.

13. When origo came into existence, it could not move, because it was countered by origo in opposing directions.

14. Origo cannot be reduced to a smallest quantity.

TRY THIS!

Realize that the 24 hours we have each day is simply one rotation of the Earth divided into 24 parts. Now look at your watch while you think about it corresponding to the rotation of the Earth. Close your eyes and try to imagine this massive sphere called Earth rotating just a little bit as you watch the time tick away. Think what the watch would tell you if it was set to measure the rotation of another planet.

CHAPTER TWO

THE ORIGIN AND DISTRIBUTION OF MATTER

In this chapter

- We will consider how matter came to be
- Be introduced to energy structures
- Consider the universal distribution of matter
- Be introduced to the inward and the outward forces
- Consider how it is that matter retains its cohesion

Outward Force

Inward Force

The First Production Run

Pre-universe, there were no laws of physics or thermodynamics; there were no governing laws of any kind, but once origo came into existence, these laws were established.

We tend to think that the laws of physics govern the universe. However, it is more accurate to say that the laws of physics are the logical manifestations of the nature of the universe.

It is the universe that determines what the laws of physics are, and not the laws of physics that determine what the universe is.

Once origo came into existence, everything that could exist within the parameters and influence of this infinite motion became possible. In addition, everything that could not exist within its parameters and influence became impossible.

At the beginning of the universe, origo was acting in all directions. This means that wherever origo was acting in one direction, origo was also acting in the opposite direction.

Origo acting in opposing directions generated extreme pressure at all points in the universe. This pressure caused origo to compress. As origo began to compress, it experienced a change of state: the first ever change of state of or in the universe.

Origo, which is infinite motion, was unable to move because of resistance from origo in opposing directions. This caused intense pressure everywhere in the universe that resulted in a quantity of origo changing from infinite motion to a stationary state.

Because the pressure was uniform throughout the universe, this change of state must have happened uniformly throughout the universe. This would have completely filled the universe with a new type of thing, something different than the origo that came into existence at the start. This new thing was matter.

This new matter was substantially different than the matter that we are familiar with, because atoms and molecules did not yet exist, and because it was in a state of flux.

As this new matter began to form, the density of origo decreased proportionally.

Because this new matter was formed by the compression of origo, a certain quantity of pressure from origo was required to hold it in this compressed state.

Because the density of origo decreased as this new matter formed, there was not sufficient pressure to hold it in this compressed state; therefore, the new matter began to decompress.

As the new matter decompressed, once again origo increased in density, thereby causing greater resistance to its movement, and therefore greater pressure throughout the universe, resulting in the recompression of matter, but on a smaller scale than the first formation.

These vacillations continued until equilibrium was finally reached between the pressure required to hold the new matter in its compressed state and the pressure generated by origo. At this point the new universe reached equilibrium, the quantity of matter per unit of volume throughout the infinite universe was set, and the speed of light was determined. The density of origo is inversely proportional to the quantity of matter in the universe.

This was the first and only time that matter would ever be generated in the universe.

That matter cannot be created or destroyed is because of the equilibrium between the pressure generated by origo and the pressure required to maintain the compression of the energy of which matter consists.

The destruction of matter would be the transformation of compressed energy back into origo. This would increase the pressure generated by origo, which would result in the compression of origo into an equal quantity of matter. The inverse of this effect would take place if matter were to be created. This equilibrium is the reason for the stability of the universe.

Energy Structures and Their Distribution

Matter, which was generated shortly after the universe came into existence, was made of compressed origo. All things made of compressed origo we will call "energy structures," because that is what they are. They are energy that has been compressed and organized into objects. We will also call all energy structures "matter," whether alone or formed into complex configurations; subatomic particles, atoms, rocks, planets, galaxies, etc., are all energy structures.

When matter first formed, origo had to exert some of its energy on it from all directions to hold it in its compressed state. Upon its formation, this matter was uniformly distributed everywhere in the universe.

The vacillations and subsequent motion of origo occurring during the formation of matter caused variations in the energy of origo coming from different directions. These variations, in turn, caused this matter to begin moving relative to other matter. This movement caused the distribution of matter throughout the universe to become slightly non-uniform. The initial motion of this matter determined the future distribution of all the matter and energy of the universe.

This matter began moving in all directions, and as it moved, it began combining and forming complex energy structures. As this process continued, these energy structures became larger and denser, forming many different types with many different properties; we call them atoms. These atoms came together to form planets, stars, and all the other energy structures with which we are familiar.

Because matter first formed uniformly throughout the universe, we can expect that complex energy structures, such as galaxies, are distributed nearly uniformly throughout the universe.

The Inward and the Outward Forces

Origo is moving in every direction from every direction; therefore, if we could see it directly, it would appear to be traveling radially inward and radially outward at all points in the universe.

We will now give names to these two different directions of origo. We will refer to origo as the inward force when it is moving toward anything, and in the case of an energy structure, until it reaches a plane ninety degrees from its direction of travel that intersects the center of the energy structure.

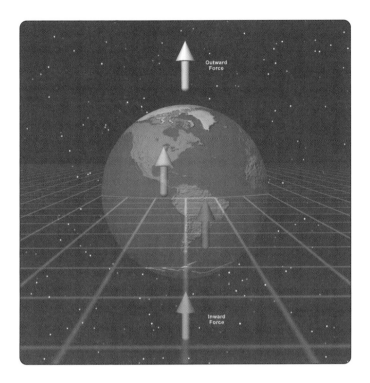

We will refer to origo as the outward force when it is moving away from anything, and with regard to energy structures in particular, when it is moving away from the above-mentioned plane.

What Holds It All Together?

The inward force acts upon a solitary energy structure equally from all directions. The inward force comes from the outside and acts upon the energy structure, always tending to move in a straight-line path.

Origo tends to move in a straight-line path because, when origo from every direction is moving in a straight-line path, resistance to its movement from each direction is equal.

If origo from one direction were to deviate from its straight-line path, it would experience greater resistance to its movement due to an increase in the quantity (or density) of origo in that location. Therefore, without some influence, the path of least resistance for the movement of origo is always in a straight-line path.

The inward force has a certain energy level before it contacts an energy structure. As it acts upon the energy structure, it loses a quantity of energy proportional to the mass of the energy structure. The energy lost to the energy structure does not just vanish without a trace; it is spent holding the energy structure together on the subatomic level and above.

Once the compression of origo produces an energy structure, in order to hold it in its compressed state, origo must continually apply a quantity of energy that is equal to that used to compress it.

If one could devise a way to temporarily eliminate the inward force on an object, the energy structure would temporarily cease to exist as an energy structure and instantly release all of the energy stored within its atoms. The energy release would occur because the energy of which the energy structure consists is compressed to a solid state; eliminating the force that maintains the compression would allow it to decompress. When the decompression takes place, the energy would expand rapidly. An energy release of this magnitude would be far greater than the amount released in a nuclear fission or fusion reaction, and the type of matter used would be irrelevant.

One could use river rock as fuel to produce an explosion that would dwarf a nuclear explosion produced by an equivalent mass of nuclear fuel. In addition, such an explosion would not likely produce the residual radiation that accompanies an atomic fission blast. This type of energy release may be the cause of bursters, massive gamma-ray bursts.

In a nuclear fission reaction, which is what most atom bombs employ, a mass of nuclear fuel, such as plutonium-239, is brought together long enough for the nuclear reaction to occur. When it begins, at least one neutron starts the reaction by striking the nucleus of another atom. This knocks several neutrons out of the nucleus of that atom. These neutrons, in turn, knock more neutrons out of the nuclei of other atoms. This chain reaction continues until great numbers of nuclei are splitting. An uncontrolled reaction such as this will produce an atomic explosion.

We know that atomic explosions are extremely powerful, but the reality is that only a fraction of the nuclear fuel actually changes into energy in an explosion such as this. The remainder changes into other elements or compounds.

DID YOU KNOW?

The first human-controlled nuclear reaction took place under the bleachers of Stagg Field at the University of Chicago on December 2, 1942, at 3:25 P.M.!

CHAPTER TWO HIGHLIGHTS

1. The laws of physics exist because of origo.

2. The universe determines what the laws of physics are, rather than the laws of physics determining what the universe is.

3. All things are possible that can exist within the parameters of origo.

4. All things are impossible that cannot exist within the parameters of origo.

5. Matter was generated by the compression of origo.

6. Continuously applied pressure from origo maintains the existence of all matter.

7. The quantity of matter in the universe is determined by the amount of pressure that can be uniformly maintained by origo.

8. Matter was generated at the beginning of the universe and will never be generated again.

9. Galaxies are distributed nearly uniformly throughout the universe.

10. The inward force is origo moving toward anything or, in the case of an energy structure, until it reaches a plane, ninety degrees from the direction of its travel that intersects the center of the energy structure.

11. The outward force is origo moving away from anything or, with regard to energy structures in particular, when it is moving away from the above-mentioned plane.

TRY THIS!

Look at any nearby solid object—your refrigerator, a rock, anything. Imagine it suddenly turning into a liquid. Now imagine it turning into a gas. Try to imagine the gas igniting and exploding into flames. Once you have this mental picture worked out, imagine the object changing from a solid form directly to the exploding gas. Look at the object again. See it for what it is, and ask yourself why this object is solid.

CHAPTER THREE

GRAVITY AND THE WARPING OF SPACE-TIME

In this chapter

- We will examine the idea that gravity causes an attraction between objects

- Consider the fabric of the universe

- Consider gravitational fields and the forces acting upon objects within and without

- Consider whether objects in a gravitational field are pulled or pushed together

- Theorize about a new type of propulsion system that could render rocket propulsion obsolete and bring about many other profound changes for our civilization

- Consider the ideas of gravitons and gravitational waves

- Realize the exact reason for inertia

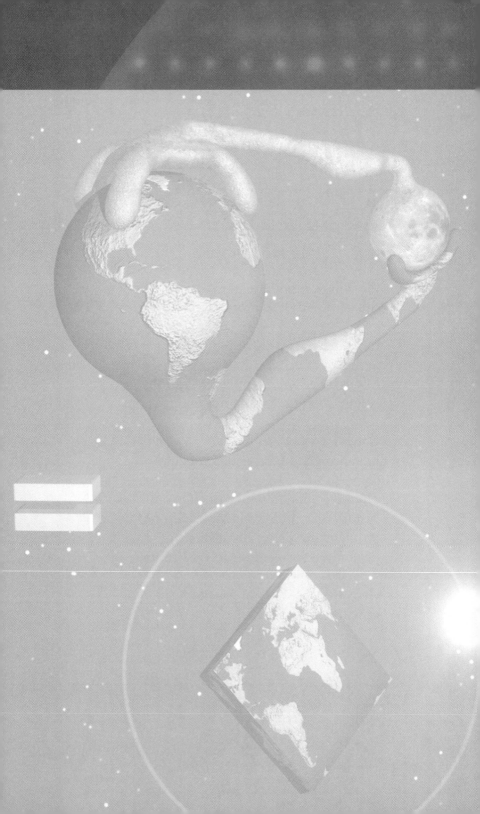

An Attraction between Objects

GRAVITY SEEMS TO BE A STRANGE KIND OF FORCE. It acts between objects that are separated by distances of many light years of seemingly empty space. "Seemingly," however, is the key word here. In chapter one, we offered the fact that gravity influences objects through space as direct evidence that space is not empty. If the space were indeed empty, there would be no gravity between objects.

Most descriptions of gravity refer to it as an "attraction between objects." Sometimes, with language such as this, the emphasis is on the fact that repulsion is not taking place. However, it seems that most of the time when people use this sort of language, the intended meaning is that the objects are somehow pulling each other together or that there is a force between them that somehow pulls them together.

An important fact we often overlook is that gravity is not even an apparent attraction between objects in the sense that they are somehow pulling each other together. Despite this obvious fact, gravity is still widely portrayed as an "attraction between objects."

To describe gravity as an attraction between objects could be correct, as long as the word "attraction" is used to mean only that objects, due to gravity, tend to move toward each other, rather than apart, and not that they are pulled together.

An observer watching an object as it falls to Earth might think that the object and Earth are pulling on each other. The observer could take all types of measurements known to man but would not find a single indication that this is what is happening.

To claim that acceleration due to gravity is because objects exert a pull on each other is perfectly acceptable within the context of a theory of gravity. However, making this claim in written curricula as if it were a proven fact is irresponsible.

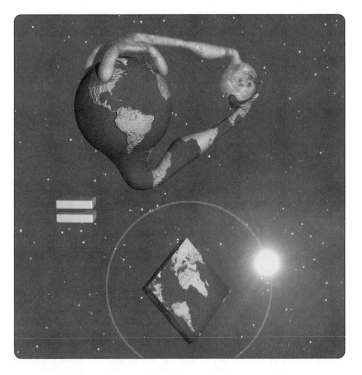

The current belief that objects pull upon one another is as ungrounded and outrageous as the archaic belief that the sun revolved around a flat Earth. Although commonly thought of and described as an attractive force, and although some may think it absurd just to ask, the question remains: is gravity the result of a pull or a push?

The Fabric of the Universe and the Gravitational Field

In his General Theory of Relativity, Albert Einstein attributes gravity to the warping of space-time due to the mass of an object. People write about the "fabric of space-time" and "the fabric of the universe" quite often but do not explain what it is. This word "fabric" means framework, structure, and a material that is woven, knitted, or felted. This is a very good word to choose for describing the universe, because we do not know what the universe consists of. Therefore, we accept the word "fabric" as a description because it gives us a feeling of something real with which we can relate.

When someone uses the term "fabric of space-time," most of us understand the idea that this person is trying to convey; however, although most of us understand the idea, what this fabric actually is remains unclear. Scientists have so far been unable to clearly describe it verbally or make an accurate drawing of it. The question remains, however; what is the fabric of space-time, and how does an object's mass "warp" it?

This theory of origo is a detailed description of an energy that fits the bill for the fabric of space-time. "Space-time," however, is not the best choice of words for labeling origo, since within the theory we consider time nothing more than the measurement and comparison of the periods of separate events.

The fabric of the universe is a better label for origo. However, the words "space-time" will be used occasionally to help the reader relate to the subject matter.

This description of origo explains how an object's mass affects this "fabric" and causes gravity. It is a description of how mass curves "space-time."

We have reasoned in chapter one that when the universe came into existence, origo was the only thing that came into existence; therefore, the universe and everything within it must be made of origo. Origo also must be the source of all energy of the universe and, therefore, the source of the force of gravity.

Although origo can deviate from a straight-line path when forced to do so, it always tends to move in straight lines. Because of this, wherever it loses energy to an energy structure such as a planet, the energy level or outward force of origo on the opposite side, as origo exits the structure, is lower than that of the energy level or inward force of origo first approaching the structure.

The difference in energy level or force between the inward and outward forces is an amount equal to that lost to the energy structure.

Because the inward force converges on the planet and the outward force radiates away from it, this region of directional energy imbalance will be nearly spherical, with the planet or other energy structure at its center. Any object within this region will experience acceleration toward the planet.

The acceleration that an object experiences within this region is equivalent to the imbalance between the inward and the outward forces, which is proportional to the mass of the bodies that are causing the imbalance.

The regions of unbalanced forces that surround all energy structures are gravitational fields; this spherical energy imbalance around each energy structure is the "curvature of space-time."

In order to relate these ideas to modern ideas of gravity, we can consider origo to be "space-time," and whenever it is diverted from its straight-line path, or otherwise in a state of imbalance, we can consider it to be warped.

The Energy Acting upon an Object Within and Without a Gravitational Field

Why acceleration due to gravity diminishes the further an object is from another is what we will consider in this section.

The inward force acting upon an object, from the direction

of and within a gravitational field other than its own, becomes stronger as the object moves further from the planet or other body at the center of the field.

As stated in the previous section, wherever origo "loses energy to an energy structure such as a planet, the energy level or outward force of origo on the opposite side, as origo exits the structure, is lower than the energy level or inward force of origo first approaching the structure."

The inward force acting upon an object from the direction of another body is weakened due to origo's passage through the other body; the inward force acting upon the object from all other directions is not weakened. This gives us a very simple way of envisioning these forces.

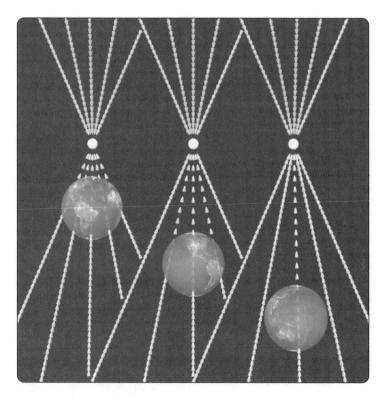

Imagine that we have a spherical object resting on the surface of another body. One hemisphere is acted upon only by "weakened" inward force from the direction of the other body, and the other hemisphere is acted upon only by "full-strength" inward force from the outside. If the object is moved a distance from the other body, these forces change.

At a distance, the "weakened" inward force that was previously acting upon one hemisphere is diluted by "full-strength" inward force coming from directions parallel to the other body's surface, thus causing the strength of the inward force from all directions to become closer to equal. This dilution is the result of the "full-strength" inward force being able to act upon the object from more directions as the object is moved further from the other body.

Picture an imaginary cone with the object at the point and the area of the other body as its base. The longer is the cone relative to the diameter of its base, the less is the acceleration due to gravity between the object and the body.

As the inward forces acting upon the object become more balanced, the acceleration due to gravity diminishes. Therefore, the further an object is from another body, the less gravity there is between them.

Although the imbalance of the inward forces acting upon the object is the result of origo losing energy to a planet, star, or other body, the body itself is not the cause of the object's acceleration. The acceleration due to gravity that an object experiences within a gravitational field other than its own is only the result of an imbalance of the inward forces acting upon the object from opposite directions.

When an object is within another body's gravitational

field, there is an imbalance of the inward forces acting upon it; therefore, there is less energy acting upon the object from one direction than from all others. If we were to remove the gravitational field, the inward forces acting upon the object from all directions would be equal. This is because the object's sides within the gravitational field that had weakened origo acting upon them would have the same strength origo acting upon them as does the rest of the object.

Therefore, an object within the gravitational field of another object has less total origo acting upon it than does the same object outside of another's gravitational field.

Is Gravity the Result of a Pull or a Push?

Is gravity the result of a pull or a push?

In the previous sections, we theorized that gravity is due to an imbalance of the inward forces acting upon an object from opposite directions. An object experiences acceleration due to gravity when the inward forces acting upon the object are less from one or more directions than from others. The object accelerates in the direction of the most reduced inward force, because the greater inward forces from the opposite directions push it toward the reduced inward force. This is the path of least resistance.

Artificial Gravitational Fields and a Revolutionary New Propulsion System

We theorized that a gravitational field is the region surrounding an energy structure in which there is an imbalance between the inward forces acting upon the energy structure and the outward forces moving away from it. We also theorized that the acceleration due to gravity that an object experiences within a gravitational field is the result of an imbalance of the inward forces acting upon the object.

If these descriptions are accurate, then the acceleration due to gravity that an object experiences is completely independent of the other body whose presence causes a gravitational field. If the body were not there, its gravitational field would not be there; however, an object's acceleration is, nevertheless, due only to the imbalance of the inward forces acting upon the object.

By utilizing these concepts of what gravity and gravitational fields are, a revolutionary new type of propulsion system could be developed in the near future. The core of this new propulsion system would be some type of electromagnetic field generator that allows the operator of a vehicle to manipulate the intensity of the inward forces acting upon all points of the vehicle.

The inward forces acting upon the vehicle at its front could be redirected and caused to act upon the vehicle at the rear, thereby causing an imbalance in the inward forces acting upon the vehicle. This imbalance will cause a forward acceleration of the vehicle that would be identical to the acceleration that the vehicle would experience in a gravitational field. We could consider this electromagnetically induced imbalance of the inward forces acting upon the vehicle to be an artificial gravitational field because it would be generated by mechanical and electrical means rather than the presence of a large body in proximity to the craft.

This new propulsion system could be used on Earth to more efficiently move ships through the water, propel submarines, and propel civilian and military vehicles, including aircraft. With this technology, the shape of aircraft will change considerably. These craft will be able to hover in place and take off and land vertically simply by reducing the inward force pushing the craft down toward the ground while increasing the inward force in the opposite direction. They will have the ability to travel at almost any speed and will be able to change direction rapidly without regard for normally encountered "G" forces. Because the inward forces acting upon the craft will be manipulated by its operator, the craft and all matter within the artificial gravitational field will exist in an inertial state independent of that surrounding the craft. This craft could be designed to be air, water, and vacuum tight, so that it could operate under water, within the atmosphere, and outside of the atmosphere.

Because origo, which is everywhere, will be the driving force behind this propulsion system, a vehicle could gain thrust in almost all environments and under almost any conditions. It may have trouble near any body that produces intense electromagnetic fields, unless those fields could either be incorporated into the propulsion system or negated.

The apparent vacuum of space will not pose a problem to this propulsion system, and it may operate better in this environment than in any other.

GRAVITY AND THE WARPING OF SPACE-TIME | 49

This document is being revised just weeks after the destruction of the space shuttle Columbia, which brings to mind the difficult maneuver of re-entry into Earth's atmosphere. This new propulsion system will make re-entry a simple and safe maneuver, removing the risks of high-velocity re-entry, atmospheric friction, and the resultant extremes of heat that pose a risk to every shuttle mission.

This technology will make rocket-powered vehicles obsolete.

The acceleration an object will experience due to an artificial gravitational field is identical to the acceleration due to a gravitational field resulting from a massive body. Therefore, a vehicle that is propelled in space by this system could reach extreme velocities. A technological advance of this type will not only make interstellar travel possible, but it will be our obvious next step and, therefore, will be unavoidable.

This type of propulsion system may enable us to travel at superluminal velocities.

This propulsion system will completely revolutionize our planet, changing everything from the way we transport people and goods to our methods of construction and destruction. However, this system will not operate on magic; a power source will be required to make it work.

The most likely abundant source of power to drive this technology will be nuclear fusion, which still has a long way to go before it will become available as a viable source of energy. On the other hand, this new propulsion system has just been conceived and development of it has not even begun, so, by the time this propulsion system is developed, it is likely that nuclear fusion will already be online. In addition, it may be that the development of this technology will play a key role in making nuclear fusion a viable source of energy. It could be that this technology will allow us to generate containment fields that will withstand the intense heat of fusion reactions.

Another option for an unlimited power supply to fuel this new propulsion system is the same technology from which the propulsion system is itself derived. This technology, which allows us to regulate the inward forces acting upon an object, may also allow us to partially block the inward force acting on an object, thereby creating a condition wherein the energy stored within its atoms can be released. If we are able to control the rate of the energy release, we would be able to convert any type of matter into energy.

A spacecraft could harvest asteroids, gases, or any other matter that could be found for use as fuel to propel it through space and to meet all of its energy needs, including the energy required to power the energy extraction system.

This would be an unlimited, non-polluting energy source that could easily provide sufficient energy for our every need, both on Earth and off.

This new propulsion system could play a key role in the long-term survival of life on Earth. If we were to detect an asteroid, comet, or other large body on a collision course with Earth, this propulsion system would give us the capability to change its trajectory, thereby causing it to pass by the Earth at a safe distance. It could also be set on a course that would take it into deep space to prevent it from returning and posing a threat again at some point in the future.

This propulsion system will allow us to build ships that can travel at extreme speeds, enabling us to go into space and intercept such incoming bodies while they are still a great distance from Earth, and it could be done quickly. With this system, decades of lead-time will not be required to divert an incoming object; the task will probably be accomplished in a matter of months, weeks, or even days. Once the ship reaches the asteroid or other body, a propulsion system similar to the one propelling the ship could be attached to the body. Once it is in place, the system could be remotely controlled to direct the body on the prescribed course.

Another option, which may be simpler but will allow direct control of the body for a shorter period of time, will be the use of large ships designed to push bodies around, much like tug boats on the water. One or more of these ships could be stationed in orbit around our planet, constantly scanning space. When a body is found to be on a collision course with Earth, the ship would go out into space to meet up with the body and steer it away from Earth.

If any bodies are made of valuable materials and are small enough to be safely recovered, the ship could slow them down, capture them, and return them safely to a processing facility on Earth.

Alternatively, any bodies that are too large to bring to the surface of the Earth could be placed in orbit around the Earth or moon for mining.

In addition to ships for planetary defense, this propulsion system will make pleasure cruises to every planet in the solar system and beyond, everyday occurrences. The stuff of popular science fiction will become everyday reality.

Gravitons and Gravitational Waves

Gravitons are the theoretical particles that many scientists believe to be the cause of gravity. The popular opinion among scientists is that these particles move in waves, similar to the way that light is believed to move. These waves are known as gravitational waves. Projects are now underway to construct instruments and facilities with which to detect and study these theoretical waves.

According to this theory of origo, the graviton is not the cause of gravity. If this theory is correct, then the graviton will never be found, because it does not exist.

According to this theory of origo, gravitational fields are due to an imbalance of the inward forces acting upon an object and the outward forces moving away from it. As energy structures change position in the universe, their gravitational fields also change position.

A binary system comprised of massive bodies such as neutron stars or black holes that are in close orbit with each other will be revolving around each other very rapidly, and their gravitational fields also will be in motion. If both bodies are in line with Earth, one behind the other, the combined gravitational field of both bodies, if it could be detected, would appear to be a single field. When these bodies revolve one-quarter turn, these two separate fields would appear to be a single field that is weaker than it was when the bodies were in line. As these bodies continue to revolve around each other, the gravity that we may be able to detect here on Earth would appear to be a single field, waxing and waning in strength. Although the effect is due only to the changing positions of both gravitational fields, this effect may be misinterpreted as being due to gravitational waves.

Inertia

Inertia is a phenomenon commonly said to be a property of matter. To refer to inertia as a property of matter is akin to claiming that gravity is an attraction between objects. Although these two phenomena are different, it is understandable how someone can arrive at these conclusions.

For example, when we observe two objects that are accelerated toward each other due to gravity, it is usually only the two objects that are seen and taken into consideration. This is understandable because it is sometimes difficult for our minds to see what our eyes cannot. However, if one chooses to see the entire picture, he will notice that there are more than just the two objects that must be considered.

Space is not empty. The electromagnetic spectrum is present throughout the universe; this alone is sufficient evidence that space is not empty. Therefore, if we choose to consider only the two objects that are being observed and exclude everything else, we make a conscious effort to ignore the rest of the picture and all other factors that may play a role in our observations.

Likewise, when considering inertia, if we choose to consider only the object that is being observed and exclude everything else, we make a conscious effort to ignore the rest of the picture and all other relevant factors.

When someone describes inertia as a property of matter,

he is defining the realm of influence, of the cause of inertia, as the volume and spatial location that is occupied by any individual object. In his shortsightedness, this man attempts to confine a universal phenomenon within the confines of finite objects.

The inward force is acting upon every object, and force, upon every aspect of the universe from all directions. If an object is at rest and the inward force is acting on it equally from all directions, some energy would be required to cause it to move. Energy would be required because in order to make the object move, the inward force acting in the direction opposite the object's motion must be countered.

The inward force acting in the direction opposite the object's motion is countered in an amount equal to the object's acceleration times its mass, per Sir Isaac Newton's second law of motion. As the inward force in this direction is countered, the inward force acting in the direction of the acceleration accelerates, also, in an amount equal to the object's acceleration.

When an object is accelerated, the balance of the inward forces that are acting upon the object from each direction is changed, but the equilibrium of the system does not change.

Things may be in a state of imbalance relative to other objects and forces; however, everything is always in equilibrium relative to the universe (the primary system). This is to say that everything is always in equilibrium relative to origo.

The inward forces acting upon an object, although not always equal from all directions, are always in equilibrium. Therefore, if an object is at rest, it will tend to remain at rest until it is accelerated by some force. Likewise, if an object is in motion, it will tend to remain in motion at the same speed and in the same direction until it is accelerated by some force.

As you can see, inertia is not a property of matter but a property of the universe that we experience through matter.

DID YOU KNOW?

When objects are accelerated in a gravitational field, field propulsion is taking place!

CHAPTER THREE HIGHLIGHTS

1. There is no evidence that supports the popular belief that objects exert a pull upon other objects.

2. The act of teaching students that objects exert a pull upon other objects is irresponsible.

3. Origo is "space-time."

4. Origo is the source of the force of gravity.

5. The outward force moving away from an energy structure is less than the inward force moving toward it because of energy spent by origo holding the energy structure together.

6. Because the inward force moving toward an energy structure is greater than the outward force moving away from it, objects will experience acceleration toward the energy structure in an amount equal to the energy that origo spends holding the energy structure together.

7. Regions of unbalanced inward and outward forces that surround all energy structures are gravitational fields.

8. The "warping of space-time" is the imbalance of, and/or the diversion of origo from its straight-line path.

9. The acceleration due to gravity that an object experiences is due only to an imbalance between the inward forces acting upon the object from opposite directions.

GRAVITY AND THE WARPING OF SPACE-TIME | 57

10. An object within another object's gravitational field has less total origo acting upon it than does the same object far removed from any gravitational fields.

11. Objects within a gravitational field are pushed together, not pulled.

12. The development of "artificial gravitational fields" that can propel vehicles in almost any environment is possible.

13. The theoretical particle called "graviton" will never be found because it does not exist.

14. Inertia is not a property of matter but is a property of the universe that is experienced through matter.

15. Everything is always in equilibrium relative to the universe (the primary system).

TRY THIS!

Go outside and look at the moon. As you look at it, realize that it is not a white disk in the sky but a massive stone sphere that is hurtling around the Earth at about 2,288 miles per hour. Now imagine how powerful the force must be that holds in orbit this gargantuan object that is 1.23% the mass of the Earth!

CHAPTER FOUR

LIGHT AND ITS MOTION THROUGH THE UNIVERSE

In this chapter

- We will consider what modern physics has to say about light when it is not being observed

- Theorize about why light travels at the speed that it does

- Consider the generation and the motion of light

- Examine several misinterpreted experiments and observations

- Consider the "Doppler Shifts" and explain why they do not indicate that the universe is expanding

- Consider how a gravitational field affects light

- Discover a "faster than light" communication system that does not violate the speed of light

- Consider the age of the universe

- Consider why the night sky is dark

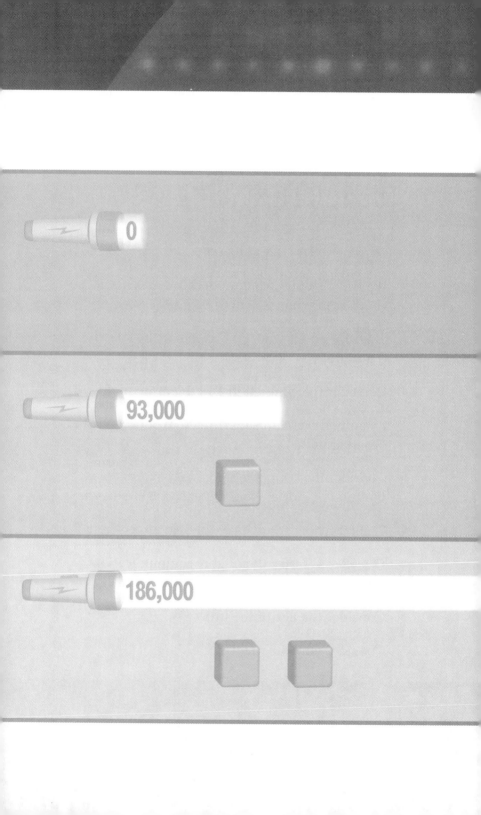

What Is Light When It Is Not Being Observed?

It has been said that this question has no meaning, as if it can never be answered. Those who answer the question in this way say that it cannot be answered by experiment or observation, because to detect or observe light negates the necessary condition of not being observed; therefore, they say it is foolish even to ask.

Imagine what it must have been like to live in the dark ages when the air itself was a mystery because it could be felt and its effects could be seen, but it could not be seen. Then look around at the modern view of light, and you will see that we are not far from the thinking of long ago.

Some experiments seem to show that light consists of tiny packets of energy called photons, while others seem to show that it consists of waves. Because different experiments can be performed that demonstrate light to apparently consist of photons, waves, or both, and because light cannot be observed until it collides with something and is absorbed and/or reflected, the reality of what light actually is when it is not being observed is left to the imagination of the experimenter.

Light is some real thing that moves in a certain manner and operates within the parameters of the laws of physics. Whether we are observing or not and whether we understand these laws or not, light still operates within these parameters. Science has no answer to the question of what light really is.

This chapter will answer the question of what light is and explain how it moves through the universe.

Why Light Travels at the Speed That It Does

Have you ever wondered why the speed of light is the value that it is? The ideas presented in this section give us a very plausible answer to this age-old question.

Light travels through the universe at a certain speed; this speed, in a vacuum, is approximately 186,000 miles per second. It seems however, that light should be able to travel at just about any speed, such as 286,000 miles per second or 186 miles per second. So why, then, is it traveling at the speed that it does?

The fact that light travels at the same speed throughout the universe seems to indicate that whatever is the cause of its speed has jurisdiction everywhere.

Although the speed of light will change when light passes through media of different densities, it does not change as time goes by. The speed of light is the same today as it was when Danish astronomer Olaus Roemer first measured it in 1676, and we have no reason to believe that it will change tomorrow. It is not 186,000 miles per second one day and 172,000 miles per second the next. Because of the constancy of its speed, scientists consider it a universal constant.

Common theories about the propagation of light claim that light travels freely through space and that it does not need a medium through which to travel. This idea assumes that what most people consider empty space is indeed empty and that light somehow propels itself through the universe.

The theory of origo does not assume that space is empty; in fact, it states that truly empty space does not exist anywhere in the universe. Some may see this theory as an attempt at reviving the long-discarded idea of all-pervading ether, through which light was once said to travel as sound waves travel through the air. This ether was supposed to be a continuous substance that existed throughout the universe and was at absolute rest.

The theory of origo is not an attempt at reviving the idea of this ether; however, it is an attempt at making sense of the universe without discounting the fact that space is not empty. Even though there is vast evidence to demonstrate that space is not empty, it seems that scientists are unwilling to fully accept this fact.

The accepted model of the atom is one whose volume scientists claim consists mostly of empty space. It seems that, according to this atomic model, many still consider space to be an empty volume within which things exist and events happen. If we accept Albert Einstein's idea that space is not a passive background in which things exist and events occur but that space and objects are inextricably connected, how can we accept the idea of empty space, even within an atom?

It is a well-known fact that light travels more slowly through denser media than it does through less dense media. If we were to watch a "beam" of light as it travels from a piece of glass into water, we would see it accelerate as it enters the less dense water. If we were to watch this same "beam" travel from the water into the air of Earth's atmosphere, we would see it accelerate as it enters the less dense air. If we continue

to watch this "beam" as it travels from the air into the vacuum of space, we would see it accelerate as it enters the less dense space. If we followed this "beam of light" through the universe until it approaches a black hole, we would see it accelerate as it moves deep into the gravitational field surrounding the black hole, within the radius of the event horizon.

This progression of light through successive media of lower densities and its subsequent acceleration is another example that demonstrates that space is not empty. If someone analyzing this "beam of light" had never before heard space described as "empty," he or she would have no reason to assume that space is anything other than a medium of lesser density than the air of Earth's atmosphere and greater density than the region surrounding the black hole, within the event horizon. How, then, is it that we can assume space to be empty?

One may claim that space is empty and that the region within the event horizon surrounding the black hole is somehow a special case. Einstein's General Theory of Relativity claims that massive bodies somehow warp "space-time" within their vicinity. If "space-time" around massive bodies is warped, whether we can directly observe it or not, there must be something there to warp. Again, if we accept this statement, how, then, is it that we can accept the idea that any space is empty?

The theory of origo claims that there is no empty space, but that the universe consists only of origo in one form or another. The following paragraphs will give further examples of this, adding weight to the idea that space is not empty and that origo propels light through the universe.

Light from two different sources will travel at the same speed while passing through equivalent media. As long as the media through which light is traveling are of equivalent density, transparency, temperature, pressure, etc., the type or intensity of the light source does not matter; light from any source will travel at the same speed. This is evidence that the source is not what propels light through the universe.

If the light source propelled light through the universe, then light from different sources would travel at different speeds, depending upon factors at the source. We know that this is not the case. Light from a firefly travels at the same speed as the light generated by a thermonuclear explosion. The source is merely the instrument generating the light.

When light enters a transparent medium of greater density than the medium that it came from, light slows down.

Light traveling through the air will slow down to a certain speed when it enters water, which is denser than air, and it will remain at this speed as long as it is traveling through water of constant density. This effect indicates that the denser medium offers resistance to the movement of light. According to the theory of origo, origo carries light through the universe, so this effect also indicates that the denser medium offers resistance to the movement of origo and that light is our visual indicator of origo's motion through the universe. This resistance slows origo's movement, as evidenced by the reduction of the speed of light.

Another very interesting fact is that when light travels from some medium into a less dense medium, it accelerates.

Light traveling through the water will accelerate to a certain speed when it leaves the water and enters the air, which is less dense than the water, and it will remain at this speed as long as it is traveling through air of the same density. This acceleration and the fact that light moves at a constant speed through any medium are very important observations that will help us to understand light.

If an object is put in motion with a one-time burst of energy and it moves through a medium offering resistance to its motion, such as water, it will eventually slow down and stop. All of the object's momentum will be transferred to the water molecules as they are displaced by the object's motion through them.

In order for the object to move through the water at a constant speed, a constant acceleration must be applied to it. This acceleration must be greater than the deceleration caused by resistance to the object's motion through the water.

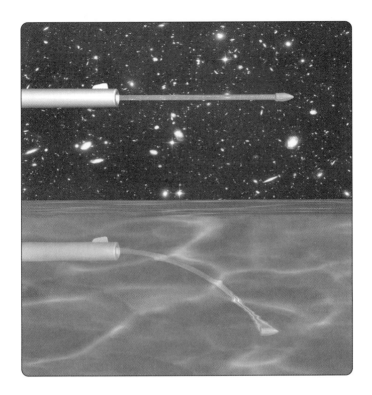

If a constant acceleration is applied to an object that is traveling through some medium, such as water, the object will reach a point where the acceleration and the deceleration balance each other. At this point, the object will move at a constant speed.

The balancing of the object's acceleration and deceleration does not mean that they are equal. If movement is taking place, the acceleration is greater than the deceleration, and the constant speed of the object is proportional to the difference between the object's acceleration and deceleration.

If the object leaves the water and enters the air, it will accelerate to a greater speed because resistance to the object's motion in the air is less than the resistance in the water, and, therefore, there is no initial balance between the resistance by the air and the object's acceleration.

As it did in the water, the object will again reach a point where the acceleration and the deceleration balance each other. At this point, the object will once again move along at a constant speed, but the speed will be greater than it was in the medium of greater density, the water.

As evidenced by the fact that light accelerates and decelerates when transitioning between media of differing densities, it is clear that light does, in fact, experience resistance to its movement. Therefore, we must conclude that, in order for light to move at a constant speed through water, air, space, and other media, it must be experiencing a constant acceleration.

This brings us back to the idea of empty space. If space were indeed empty, then light would experience no resistance to its movement, and if light were experiencing a constant acceleration with no resistance, it would accelerate to an infinite speed. The fact that light travels through empty space at a certain speed less than infinite is further evidence that space is not empty, and the fact that light experiences a constant acceleration is evidence that it is propelled by something that has constant motion throughout the universe, something other than itself or its source.

In chapter two, we theorized that when origo came into existence, due to resistance from origo acting in opposing directions, movement could not take place. In order for it to move, a change of state had to occur. As origo acted in opposing directions, pressure was generated and it began to compress. Due to this compression, some origo changed state. The result of this compression was the formation of matter.

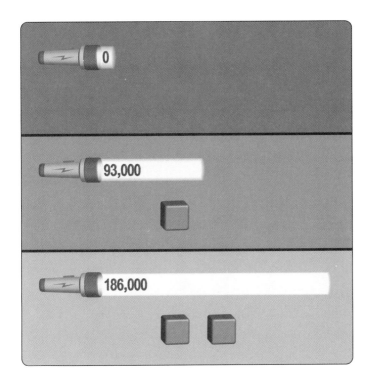

As matter formed, the density of origo decreased proportionally. The reduction of the density of origo allowed it some movement, in an amount equal to the speed of light. Therefore, the quantity of matter in the universe is inversely proportional to the density of origo, and the speed of light is proportional to the quantity of matter in the universe.

The above statement seems to imply that origo is able to travel at the speed of light, which is an inaccurate implication. The theory of origo claims that light is carried through the universe by origo; therefore, it is more accurate to say that light travels at the speed of origo.

If, when matter formed, the density of origo could have decreased more than it did, there would now be more matter in the universe than there is, and the speed of light would be faster than it is. In addition, if, when matter formed, the density of origo could have decreased less than it did, there would now be less matter in the universe than there is, and the speed of light would be slower than it is. It is the density of origo that determines its speed; less density allows greater speed, while more density allows lesser speed.

We reasoned in chapter one that origo came into existence with infinite velocity; if for example fifty percent of the total quantity of origo per unit of volume had compressed into matter, it seems that the remaining origo, and therefore light, would be traveling at an extreme speed relative to the cosmic scale. The speed of light seems like an extreme speed from our earthbound viewpoint; however, it is a snail's pace when viewed on the galactic scale, and when viewed on the cosmic scale, it is almost motionless. Almost motionless is not what we would expect if a large percentage of origo with infinite velocity had compressed into matter. Therefore, it is safe to conclude that only a small percentage of the total quantity of origo per unit of volume remained compressed as matter after the universe stabilized.

The Generation of Light

It seems that modern science does not fully understand light and its motion through the universe. When we finally do understand light, we will find that existing experiments and observations showed us its secrets all along, but we simply did not see them. Perhaps all we needed was a little more illumination.

LIGHT AND ITS MOTION THROUGH THE UNIVERSE | 71

Light is a form of energy, moving at a different speed and in a different manner than other forms of energy. For example, heat is energy, and a hot object will radiate both heat and light, but this is only true when the heat reaches the surface of the object. When the heat is only on the inside of the object, light does not radiate away from it. The heat within the object does not travel at the speed of light; it moves at a much slower pace as the energy transfers from atom to atom.

Forms of energy include kinetic energy, gravitational potential energy, deformational energy, static electricity, and more. Although some forms of energy do not travel at the speed of light, all forms can be converted into light.

Let us consider how light is generated. When a sufficient quantity of energy is applied to the atoms of a material, such as the tungsten filament of an electric light bulb, visible light is generated.

If the atoms of the filament were at a temperature of absolute zero, they would be at their lowest energy level and would not emit any radiation. If energy was to be applied to these atoms and their temperature was raised, even by a few degrees, this added energy would radiate outward away from the filament. If enough energy were applied to the atoms of the filament, the radiation emitted would be in the form of visible light.

When electricity is applied to the atoms that make up the filament, they realize an increase in energy. This energy increase raises the energy of the atoms to a level greater than their normal level. The atoms of the filament cannot contain the extra energy that is continually being applied to them; it must be released.

The first law of thermodynamics states that in any interaction, the total amount of energy does not change. This means that the extra energy that is being released by the filament's atoms cannot just vanish.

There are three factors to consider about the extra energy being applied to the filament's atoms: how much energy is being applied in a given period, how much energy the filament's atoms can release in a given period, and how much of this extra energy can be accepted in a given period by atoms adjacent to the filament.

If atoms adjacent to those of the filament, due to their individual properties, cannot accept energy from the filament at the rate at which it is emitted from the filament, then there is a surplus of energy coming from the filament. This surplus energy cannot go back into the filament, because that would be the path of greatest resistance.

The path of least resistance for the dispersion of this surplus energy is outward, away from the filament. When an energy surplus exists, this energy is released to the only thing capable of accepting it, origo. When it is released, origo picks it up and it begins to travel with origo, at the speed of light.

The energy as it existed within the atoms of the filament was not part of the electromagnetic spectrum, but as soon as it was released to origo, it became part of the electromagnetic spectrum. That is how light is generated.

If more energy per unit of time is applied to the atoms of the filament than they are able to release to adjacent atoms or to origo in an equal unit of time, the atoms will experience a change of state to one that allows them to release energy at the rate that it is being supplied to them. This is why materials undergo changes of state such as solid to liquid and liquid to gases, etc.

How does light energy begin to travel with origo? As stated in chapter one, origo travels in all directions everywhere in the universe. When the extra energy in the atoms of the filament is released to origo, the energy level of origo at that point is increased to a level above its normal state. This energy increase is realized by origo traveling away from the atoms in every direction, the outward force.

Because origo realizes the energy increase and carries the surplus energy away in every outward direction, its motion is that of an expanding sphere, a radially outward motion. This is why, when you turn a light on in your house, the light seems to illuminate all sides of the room at once.

The Frequency of Light

When the atoms of the filament release energy to origo, what happens to the origo that realizes the energy increase?

For every action, there is an equal and opposite reaction. Therefore, origo, traveling along at the speed of light, must experience some change as it suddenly expends energy while accelerating a quantity of energy that had been traveling at sub-light speeds within the filament to light speed outside of it.

It may be said that the energy within the filament was electricity and, therefore, was already acting at the speed of light; however, this is not the case. The electricity in the wires leading to the filament may be traveling near the speed of light, but within the filament, this energy is converted to heat, which does not travel at the speed of light. This heat makes up the surplus energy that is released to origo.

For origo to accelerate a quantity of energy to the speed of light, it must lose some energy equal to that gained by the energy it is accelerating. Therefore, the greater the light energy increase realized at a point by origo, the more energy that origo must transfer to the energy that it is accelerating. This results in origo slowing down momentarily at that point.

Origo does not slow down permanently; the amount of deceleration that origo realizes is equal to the momentum of the newly accelerated energy. This effect of origo momentarily slowing down as it accelerates a quantity of energy to the speed of light will appear as a certain frequency or wavelength to someone analyzing the light.

The quantity of energy released to origo per unit of time, which will be a result of the type of atoms releasing the energy and the amount of energy supplied to them, will determine the observed frequency.

The Motion of Light

As the atoms of a material release surplus energy to origo, that energy begins to travel radially outward with origo at the speed of light; it has become light or some part of the electromagnetic spectrum. This light energy makes an expanding sphere, lengthening its radius at the speed of light. As the light travels and expands radially, the quantity of energy per unit of area at the leading edge of the sphere decreases in proportion to its expansion.

LIGHT AND ITS MOTION THROUGH THE UNIVERSE | 75

We will now look at the leading edge of this expanding sphere of light, which we will refer to as an "instance of origo."

We will attach the label "instance of origo" to any leading edge of origo that is being taken into consideration, whether it be the inward force, outward force, or other, and whether it carries light energy or not.

The light traveling radially outward from a source is not propelling itself through the universe; origo carries it along. Because origo is traveling in and from every direction, the light traveling radially outward with origo also has origo traveling toward it in the opposite direction, the inward force.

The effect that the inward force has on the light energy is that, as this instance of origo expands, the inward force places pressure on the light energy, causing it to tend to spread out equally around the instance of origo. This effect is similar to the way that the oceans of Earth tend to spread out equally due to the Earth's gravitational field, which is due to the same exact force. This means that light not only travels in a straight-line path away from its source, but also travels laterally around the instance of origo that carries it.

How does light travel laterally? Light cannot exceed the speed of the origo that is carrying it. Light must travel with origo and cannot slow down and escape from the instance of origo without colliding with and transferring its energy to something else. There is, however, nothing to prevent lateral movement around the instance of origo.

Light moving laterally in an instance of origo does not change its straight-line speed because origo, which is carrying it, continues to move relentlessly outward at a constant rate away from the source. Also, the light energy does not need to make any sort of "jump" to change its radially outward direction as it moves laterally in an instance of origo, because origo is moving in an infinite quantity of directions. Therefore, it is a smooth and seamless transition from one direction to another.

All surplus energy released by energy structures to origo becomes energy traveling at the speed of light. This includes the entire electromagnetic spectrum.

Imagine a spherical instance of origo carrying light as it overtakes and passes a planet. Before it reaches the planet, it has a uniform energy level all around it. As it reaches the planet, the light energy that it is carrying is absorbed and/or reflected by the planet. And the portion of the instance of origo that contacts the planet loses a quantity of its energy to the planet, equal to the mass that it passes through.

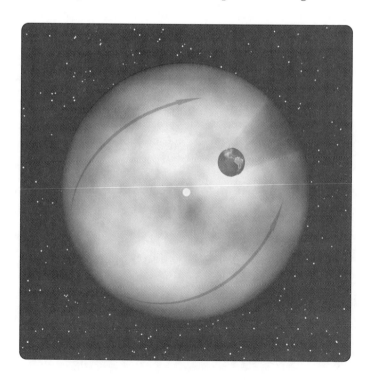

The instance of origo, after part of it passed through the planet, is no longer of uniform energy level all around. The origo that passed through the planet now has less energy than does the rest of this instance of origo and it has lost its light energy.

As was stated earlier in this chapter, "the inward force causes pressure on the light energy, which causes it to tend to spread out equally around the instance of origo." This pressure on the light energy, due to the inward force, causes the remaining light energy in this instance of origo to tend to equalize. In order for it to equalize, the light energy in the area that passed through the planet must increase. The only way that this can increase is through the lateral movement of the remaining light energy in this instance of origo into this low-energy area. As the energy begins to equalize, the level of the light energy throughout this instance of origo decreases. This effect happens every time an instance of origo carrying light passes through an energy structure.

This explanation of what light is and how it travels through the universe is very different from the commonly accepted ideas, and although scientists base the accepted ideas about light upon observations and experiments, they are misinterpreting these observations and the results of these experiments.

Misinterpreted Experiments

In the previous section, reference was made to certain experiments and observations whose results have been misinterpreted. These experiments and observations are as follows:

- The Michelson-Morley experiment
- The apparent wavelike property of light, as shown by experiments that produce interference patterns
- The apparent corpuscular nature of light, as shown by the photoelectric effect and how light affects a photographic plate
- Red- and blue-shifted light from other galaxies

Modern physics and cosmology are largely founded on the interpretations of these observations and the results of these experiments. The implications of these misinterpretations are profound.

There are a great many puzzles that modern physics cannot supply an answer for, and many answers that are far from satisfactory. There is obviously some information missing, and there is misinformation that is not allowing the correct answers to become evident. Many will tend to believe that it is mostly a case of missing information that science has yet to discover.

It is apparent that modern physics and cosmology are relying on misinformation from the interpretations of certain experiments and observations, thereby arriving at an inaccurate interpretation of the universe. This misinformation is preventing answers to many mysteries of the universe.

Ask yourself the following question and answer it truthfully, no matter how improbable it may seem to you. Is there any possibility that the results of the observations and experiments listed above may have been misinterpreted?

If you answer in the negative, you might as well stop reading right now and burn this book. However, if you answer in the positive, read on.

The Michelson-Morley Experiment

The Michelson-Morley experiment was designed with the reasoning that light is a wave. In the Michelson-Morley experiment, an apparatus is built in which a single "beam" of light is split, half of the original beam reflected ninety degrees away from the other half of the beam and then made to meet up with it again. As the apparatus moves through the universe with the rotation of the Earth on its axis, its revolution around the sun, etc., the two beams of light will move through the universe with varying speed and direction relative to each other. Because light is apparently a wave, as shown by James Clerk Maxwell in 1864, the two beams should arrive out of phase with each other and cause interference with each other when they merge.

The idea was that, if interference between the two beams occurred, this would prove that light traveled at a fixed speed through the universe. However, no interference occurred; therefore, the interpretation reached was that because light is a wave and interference did not occur, the motion of the body (apparatus) relative to the ether produces a contraction of the body in the direction of motion that makes it appear to be the same speed in every direction. Albert Einstein's interpretation of this experiment is that all freely moving observers (including the separate parts of the apparatus) should measure the same speed for light, no matter how fast or in what direction they are moving.

According to the ideas expressed in the theory of origo, light is not a wave, even though some experiments make it appear to be.

The result of the Michelson-Morley experiment could have been interpreted in this way: "the fact that no interference pattern occurred with this experiment indicates that light is not a wave, and all interpretations of previous experimental results that seem to indicate that light is a wave must be re-evaluated."

The fact that no interference pattern occurred with the Michelson-Morley experiment clearly shows us that light is not a wave.

As light travels radially outward with origo, the leading edge of the expanding sphere of light energy can be thought of as a wave front; if two of these collide, the energy from each will interact with the other. In fact, the light energy in a single instance of origo can be disrupted and caused to interact with itself as it moves laterally. When an object such as the material that makes up the space between the slits in a double slit experiment disrupts the light energy in an instance of origo, it interacts with itself as if the energy were from two separate sources, thus producing the typical interference pattern.

As light energy moves laterally in an instance of origo, there may be greater concentrations of it in some areas than in others. These areas may form regular patterns that may be interpreted as being waves. However, it is not correct to describe the light energy within these waves as consisting of or being waves without describing what the waves are comprised of.

Light energy is completely fluid, and may have greater concentrations in one area than in another, but the energy itself is not a wave. A good example of this is water. We can detect and measure water waves on the ocean, and can gather much data about them, but we do not refer to the water, of which these waves consist, as waves. Likewise, we should not refer to the energy making up light waves as waves. We should call it by what it is: light energy in the form of origo waves.

Experiments using a diffraction grating, double slit experiments, and others show exactly how light travels through the universe in a radially outward direction, and how the light energy in an instance of origo will move laterally around the instance of origo, due to the inward forces acting upon it. These types of experiments serve very well in confirming the lateral motion of light in an instance of origo.

The Apparent Corpuscular Property of Light

The photoelectric effect and the effect that light has on a photographic plate show the apparent corpuscular nature of light. This alleged property of light is also a misinterpretation of accurately measured and observed events.

Certain experiments demonstrate that electrons are emitted when the right type of light strikes the right kind of metal. Light has also been shown to leave in discreet packages of energy called photons when it is emitted. The experiments that demonstrate these effects investigate how light interacts with the atoms that make up materials.

All experiments that show the apparent corpuscular nature of light employ light acting directly with the materials to obtain results. Experiments that are said to demonstrate the apparent corpuscular nature of light tell us more about the nature of atoms and how atoms absorb and emit light energy than they tell us about the nature of light itself.

According to the theory of origo, light consists of completely fluid energy being carried along by origo; it does not exist in "packets" as the idea of the photon suggests. Evidence of the existence of photons is derived from experiments that show photoelectric emission and from those that measure the energy decrease in a material as it gives off light.

For a material to absorb energy from light, the light must have sufficient energy to penetrate the atoms of the material and thereby increase their energy levels. This happens only at certain energy levels, and the properties of the material's atoms determine the amount of energy required for penetration. If the light striking the material does not have sufficient energy, it will not penetrate the atoms; if it does have sufficient energy, the light penetrates the atoms and they realize an increase in energy, at which time photoelectric emission may take place.

Experiments that show the energy drop as a material gives off light are accurate but misinterpreted. Experiments show that a tiny packet of light is given off when what is called an atomic oscillator drops down one energy step. The misinterpretation is the idea that a tiny packet of light, a photon, is given off. What is actually happening is that a tiny quantity of energy is given off.

This released energy increases the energy level of origo and becomes light as origo carries it away, but the tiny packet of energy given off by the atoms is not the light, it is only energy that becomes light as it leaves the atoms.

Just as atoms can receive energy only at certain levels, atoms can release energy only at certain levels; the reason for this will be explored in chapter six.

LIGHT AND ITS MOTION THROUGH THE UNIVERSE | 83

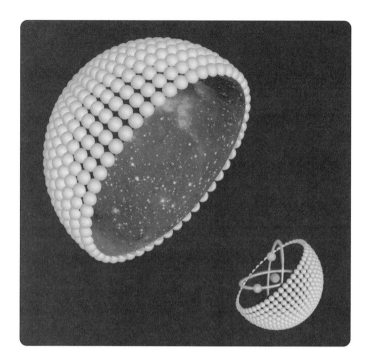

Based upon this explanation, photons do not exist as "particles" of light. A photon is merely a measurement of the quantity of energy required for an atom to absorb and emit light energy.

In order for us to see a distant galaxy, photons from every unobstructed light source must reach our eyes. Each individual atom at each star's surface emits the alleged photons given off by the stars.

Because the galaxy can be seen from any location indicates that each individual atom was emitting photons radially outward. Any observer with an unobstructed view of a photon-emitting atom should be able to see it at any point.

As an observer changes location relative to the distant galaxy, the image of the galaxy does not come and go; this is evidence that there are no gaps between the alleged photons.

One might say that it is not the light from individual atoms that the observer is seeing, but light from the galaxy as a whole. If this were the case, the observer would not see any definition, but would see only a single bright area. The fact that we can see details of distant galaxies is evidence that the light that we see comes from individual atoms.

If light actually exists as photons, how can the same quantity of photons that occupied the surface area of a single atom also completely occupy an area that is equal to the surface of a sphere that has a radius of light years or even billions of light years?

According to the theory of origo, light does not exist as photons. The reason that we can observe details of distant objects without the image coming and going is because light is completely fluid energy.

The Diffraction-Box Experiment

The diffraction-box experiment is an experiment in which a single photon allegedly travels simultaneously through two slits to form an interference pattern on a photographic plate at the end of a long box. This is an experiment that clearly demonstrates that "photons" do not exist as particles of light, but are merely measurements of the energies required for an atom to absorb and emit light energy. The single photon that is said to be in the box is not a single "particle" but is instead a quantity of completely fluid energy in the form of light that is equal to the minimum energy required for a single emission from the atoms of the material used to produce the light.

A single "particle" or "photon" does not go through two slits at the same time; this is illogical. Origo carries the completely fluid energy, which was released to it by the light source, radially outward in all directions. Energy goes through both slits, but it is not in the form of a particle; it is completely fluid energy that is everywhere in the box.

Light consists of completely fluid energy that is carried by origo.

As evidenced by the diffraction-box experiment, light does not consist of photons.

Edwin Hubble's Doppler Shifts (I)

In 1912, American astronomer Vesto M. Slipher discovered that the line spectra of radiation emitted by other galaxies were shifted toward the red end of the spectrum. American astronomer Edwin Powell Hubble interpreted this as being due to the Doppler effect, which meant that almost all galaxies are moving away from one another. This interpretation led to the conclusion that the universe is expanding.

After more than a decade of measuring Doppler shifts, cosmologists determined that galaxies at different directions but equivalent distances from Earth had equivalent Doppler shifts. They soon concluded that this relationship between speed and distance was evidence that the universe is expanding uniformly. Not only do they say that the universe is expanding, but also that it is expanding at an ever-increasing rate the further the distance from Earth.

It is currently believed that not all of the spectral shifting is due to the Doppler effect but that it is also due to the "fabric of the universe" itself, expanding. This type of expansion is called cosmological expansion.

If the idea of cosmological expansion is correct, at some distance, objects will be moving faster than the speed of light; therefore, light from these objects could not reach Earth. This distance is called the Hubble radius.

According to the theory of origo, the accepted interpretation of the red- and blue-shifted light from other galaxies is incorrect.

The interpretation of the so-called Doppler shifts, according to the theory of origo, is as follows: when a quantity of light energy in an instance of origo is removed from the instance of origo, the remaining light energy must move laterally around the instance of origo in the direction of the loss. This movement is due to the inward forces acting upon the light energy as it expands radially outward.

Imagine the surface of the Earth, smooth and completely covered with water. If one were to take a bucket, dip it in and remove a bucketful of water, water from everywhere on the planet's surface would begin to flow laterally into this area. This flow would be due to gravity causing pressure on the water at all locations. Due to this pressure, the water level over the entire surface of the planet would tend to equalize.

This lateral flow of water toward the area where water was removed by the bucket would happen around the entire circumference of the planet, until the water level had again equalized.

For this example, the water represents the light energy. The sphere that the water occupies represents the instance of

origo. The bucketful of water that is removed represents light energy that has been removed from the instance of origo. The pressure on the water due to gravity represents the pressure on the light energy in the instance of origo, due to the inward forces acting upon it. Both pressures are the same exact forces. The lateral flow of water around the planet represents the lateral flow of light energy around the instance of origo, and the water level represents the level of light energy throughout the instance of origo.

An instance of origo carrying light from a distant galaxy will have many "buckets of water" removed. The further the galaxy is from us, the larger the instance of origo and the more "buckets of water" that have been removed by other galaxies, which translates into more lateral flow of light energy in directions away from us.

This flow of light energy in lateral directions away from us is what causes the red shift, and the greater the flow, the greater the red shift. The further the galaxy is from us, the greater the red shift will be. This effect was confirmed experimentally in 1960 by Robert Pound and Glen Rebka at Harvard University.

The flow of light energy in lateral directions away from us explains why the red shift is greater the further the light source is from us; it is not due to the Doppler effect or cosmological expansion, though this is the widely held belief.

According to the theory of origo, light does not consist of rays, waves, or particles; origo carries completely fluid light energy radially outward from its source. Therefore, if we can see light from a galaxy that is ten billion light years distant, there is a sphere of light energy at least twenty billion light years in diameter that must be taken into consideration.

When light energy is removed from an instance of origo by reflection, absorption, or other means, the remaining light energy is forced laterally around the instance of origo by the pressure of the inward forces, and so it tends to equalize. This means that if light energy anywhere in this instance of origo is lost, the effects of this loss are realized throughout the instance of origo. On the opposite side of the instance of origo, twenty billion light years away, the remaining energy moves laterally in the direction of the loss.

The Red-Shift Communication System

The idea of the lateral motion of light energy due to a loss of it somewhere in an instance of origo seems to allow for near-instantaneous communication at great distances. For example, a signal source being used for red-shift communication would be positioned halfway between communication stations. This signal source would always be energized, sending light or some other electromagnetic signal radially outward in all directions with origo (it may be possible to use stars as signal sources). This signal would be monitored continually at all stations. A Morse code or other type of message could be sent almost instantaneously by somehow removing a certain quantity of electromagnetic energy from the instance of origo, whose radius intersects separate communication stations.

Removing this energy would cause a red shift in the energy received by the other communication stations. If this energy removal followed a pattern, such as Morse code, the red shift in the energy received at other stations would come and go in that same pattern.

Although one could send and receive information across distances equal to light minutes, hours, and possibly years almost instantly, no component of the red-shift communication system would travel faster than light.

In the red-shift communication system, the signal source is always energized; therefore, the signal is already spanning the distance between stations. The signal is already sent and received before information is sent. The most difficult part of implementing a system such as this will be removing enough energy from the instance of origo to cause a measurable red shift at the other stations.

Although currently accepted theories exclude superluminal communication, as the theory of origo becomes understood, it will become clear that superluminal communication is indeed possible.

In addition to introducing new technologies, the concepts of the theory of origo that are set out in this chapter will completely and logically explain what is now known as "quantum entanglement."

How a Gravitational Field Affects Light

Now let us consider another effect that energy structures have on light.

As an instance of origo carrying light with a uniform energy level approaches a gravitational field, the inward force on the instance of origo is reduced from the direction of the gravitational field. This causes the light energy to be forced laterally around the instance of origo in the direction of the gravitational field, as it tends to equalize due to the inward forces acting upon it.

For example, imagine that the surface of the Earth is smooth and completely covered with water. Now imagine that gravity is eliminated at one point on the Earth, such as at one of the poles. The result would be that gravity would force all of the water around the planet and it would leave the planet at this pole. The difference between water and light in this analogy is that light cannot leave the instance of origo that carries it; instead, the light will compress.

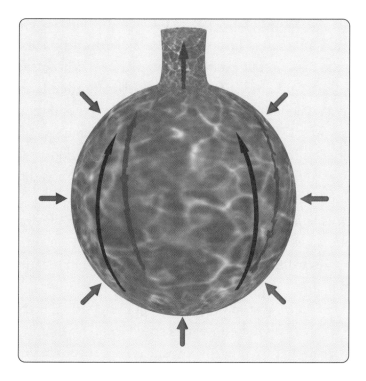

When this effect is taken into account, we see that when an instance of origo carrying light passes through a planet, the quantity of light energy removed from the instance of origo due to absorption and reflection by the planet is greater than it would be if this compression did not take place. The light energy per unit of area that collides with the planet is greater than the light energy per unit of area throughout the rest of that same instance of origo.

Any light energy that passes near to, but does not collide with the body whose gravitational field is causing the diversion of origo will curve around the body.

A phenomenon called gravitational lensing is a direct result of the above-described effect. Origo carrying light energy or other electromagnetic spectra through the universe is deviated from its straight-line course, due to the reduction of origo from the direction of every gravitational field. This deviation causes the light energy to appear to originate from locations not in a straight-line path between the observer and the actual location of the source. Probably the most famous example of this effect is the results of the 1919 British expeditions to Brazil and West Africa, in which the "bending of light" was directly observed during a total solar eclipse.

Does light carried by origo, traveling toward a gravitational field, accelerate above its accepted speed in a vacuum, due to less inward force coming from the direction of the gravitational field?

If resistance due to origo traveling in opposing directions is what causes origo to travel at a speed less than infinity, and if the quantity of matter in the universe is inversely proportional to the density of origo, then it stands to reason that if origo were less dense it would travel faster throughout the universe, and the speed of light would be greater than 186,000 miles per second. If this is the case, then, logically, if the density of origo is reduced from one direction, the speed of origo from the other direction must and will increase. Therefore, origo traveling toward a gravitational field must accelerate to a greater speed than normal, due to and equal to the reduction of origo coming from the opposite direction.

The speed of light will be greater in directions toward a gravitational field than in other directions.

Edwin Hubble's Doppler Shifts (II)

An instance of origo carrying light from a galaxy very near to us, such as the Andromeda galaxy that has not lost many "buckets of water" will not have light traveling away from us laterally. The pressure of the inward force on the instance of origo from the Andromeda galaxy will be reduced from the direction of our Milky Way galaxy because of its gravitational field. This will cause the light energy from Andromeda or another nearby galaxy to travel laterally in directions toward us. This flow of light in lateral directions toward us is what causes the blue shift.

By this definition, the red and blue shifts are not Doppler shifts. If the red- and blue-shifted light from other galaxies is not due to the Doppler effect, then the universe is not necessarily expanding but may instead be static.

In addition, rather than indicating the velocity of a star or galaxy, the measured red- and blue-shift values indicate the quantity of matter in a given volume.

To determine the quantity of matter in a given volume, we must first determine the blue-shift value of the nearest galaxy, Andromeda, in order to determine how much of an effect our galaxy's mass has on the measured red-shift values from distant galaxies. With this information we can adjust red-shift values to compensate for the Milky Way's mass.

Next, we must determine the red-shift value of the nearest red-shifted galaxy.

Then, we must calculate the quantity of matter within the relevant sphere surrounding this galaxy. The volume of the relevant sphere can be calculated by first determining the distance to the light source; this distance is the radius. Knowing this radius, we can easily calculate the volume of the sphere of which it is a part. A comparison can be made between this observed quantity of matter and the red-shift value. This will give us a ratio that will enable us to calculate the quantity of matter in any given volume, once we determine the central star or galaxy's red-shift value and the volume of the relevant sphere being measured.

Because galaxies are distributed fairly uniformly throughout the universe, the amount of matter within each equivalent relevant sphere will be about the same. This means that the observed red shift of equidistant sources will be about the same.

The Age of the Universe

If the red- and blue-shifted light from other galaxies is not due to Doppler shifts, then the estimated age of the universe could be off by any amount. The age of the universe is currently based on observing the supposed rate of expansion of galaxies, estimating the size of the universe and extrapolating the time that they all would have been at the same point in space.

If the universe is found to be not expanding, then most likely the big bang never occurred.

According to the theory of origo, the universe is static, is infinite in size, and came into existence everywhere at the same time.

Why the Night Sky Is Dark

The common belief is that if the universe is static, not expanding or contracting, and infinite in all directions, as the theory of origo claims, every line of sight would end at a star, and the night sky would be as bright as the sun. Based on the darkness of the night sky, the common view implies that somehow the stars flared to life a finite time ago and that the light from distant stars has not yet had time to reach us. This line of reasoning makes sense as long as the light energy coming from all of the stars travels only in a straight-line path between the star and Earth and reaches us at full strength.

We reasoned earlier that "if we can see light from a galaxy that is ten billion light years distant, there is a sphere of light at least twenty billion light years in diameter that must be taken into consideration." When light energy is removed from this sphere by reflection, absorption, or other means as it travels between its source and Earth, the remaining light energy is caused to move laterally toward the loss, thereby diminishing the level of light energy throughout the sphere of light. The level of light energy in an instance of origo diminishes as it travels laterally to equalize after a loss. In addition, this energy level diminishes as the size of the sphere increases. Therefore, there is a limit to the distance from a source at which light will still be visible to the naked eye.

The distance at which light from a source will still be visible to the naked eye depends upon certain factors and varies, depending upon the light source and the location of the observer. The absolute magnitude or brightness of the light source and the quantity of matter within the relevant sphere of light under consideration are two of the factors that determine the distance at which light from a source ceases to be visible to the naked eye. The relevant sphere of light has the source at its center, and its radius is equal to the distance between the source and the observer.

Because galaxies are distributed fairly uniformly throughout the universe, the distance from a source at which light ceases to be visible to the naked eye should be approximately the same for all sources of the same absolute magnitude, because the amount of matter within each relevant sphere of light would be about the same. This means that stars at a certain distance from the Earth could have been shining for eternity (although we know that they have an actual shelf life) and that light from them would still never have been visible to the naked eye from Earth.

By this explanation, we can see that the fact that the night sky is dark does not necessarily mean that it is this way simply because light from distant stars has not yet had time to reach us. Even though (according to the theory of origo) almost every line of sight does end at a star, the night sky is dark because the level of light energy within each instance of origo from distant sources is diminished to a level that is too faint to be detected by the naked eye.

DID YOU KNOW?

It takes light about 100,000 years to travel from one edge of our galaxy to the other!

LIGHT AND ITS MOTION THROUGH THE UNIVERSE | 97

CHAPTER FOUR HIGHLIGHTS

1. Origo carries light through the universe.

2. Light experiences resistance to its movement through the universe.

3. Light experiences constant acceleration through the universe.

4. Light travels at the speed of origo.

5. The speed of light is proportional to the quantity of matter in the universe.

6. The speed of light is almost motionless on the cosmic scale.

7. Light is surplus transient energy that is released to and carried by origo.

8. Light travels radially outward from its source because it is carried away by the outward force.

9. Light frequency is caused by origo temporarily slowing down as it accelerates a quantity of energy to the speed of light.

10. The Michelson-Morley experiment clearly shows us that light is not a wave.

11. Light does not exist as photons.

12. Experiments that are said to demonstrate the apparent corpuscular nature of light tell us about the nature of atoms rather than the nature of light.

13. A "photon" is a quantity of energy that is equal to the minimum energy required for a single emission from the atoms of the material used to produce the light; it is not a particle.

14. Light is completely fluid energy.

15. Red- and blue-shifted light is not due to the Doppler effect or cosmological expansion.

16. A superluminal communication system can be developed that uses red-shifted light as the signal.

17. Measured red- and blue-shift values indicate the quantity of matter in a given volume.

18. The universe is relatively static and is not expanding or contracting.

TRY THIS!

At night when the moon is bright but not full, look up at it and realize that the bright curve is pointing at the sun. You can often use the moon to highlight the sun's position, even when you can't see the sun.

CHAPTER FIVE

BLACK HOLES AND MATTER AT THE SPEED OF LIGHT

In this chapter

- We will consider the escape velocity at a black hole
- Consider an event horizon
- Consider what happens to matter near the surface of a black hole
- Consider the alleged speed limit: light speed
- Consider the possibility of matter exceeding the speed of light

What a Black Hole Is

A black hole is an energy structure of such great density that all of the energy that collides with it becomes part of it. Light does not reflect off or radiate from its surface.

Black holes have extreme gravitational fields, so strong that any matter that is caught in the gravitational field of a black hole never reaches the surface as atomic matter; all of its atoms are converted to electromagnetic energy before they reach the surface. However, the inward forces acting upon the black hole compress this and all other such electromagnetic energy into homogeneous matter, once it reaches the surface of the black hole.

Escape Velocity

Why is it that light cannot escape from a black hole?

The common idea is that the gravitational field of a black hole is so strong that the escape velocity is greater than the speed of light and that any light emitted by the black hole is pulled back toward it. This idea allows for what is known as an event horizon.

An event horizon is an area surrounding a black hole at a distance, separating the region from which light can escape and the region from which it cannot. It is believed that light will hover at this point, never escaping and never falling back toward the black hole.

Most illustrations of this event horizon depict light traveling in a straight-line path away from the center of the black hole. Such a concept suggests that light travels under its own power and can propel itself indefinitely.

This idea differs considerably from the concepts described in origo theory, which credits a force called origo for the existence of the universe and everything that happens in it, including the propagation of light.

Using the concept of origo, we will now answer the question of why light cannot escape from a black hole.

As was stated in chapter two, the amount of energy that origo loses to an energy structure is equal to the mass of the energy structure. In addition, the amount of energy that origo loses per unit of volume to an energy structure is equal to the energy structure's density. A black hole is as dense as an energy structure can possibly be. Origo loses all of its energy to a black hole; therefore, its gravitational field is also as strong as a gravitational field can possibly be. This means that origo is moving toward a black hole from every direction, but is not moving directly away from it at all.

If light were generated near the surface of a black hole, it would not be carried away by origo in every direction, as would normally occur in a region without a black hole in the vicinity. Because origo is not moving away from the black hole at all, any light generated near the black hole's surface can only be carried by origo parallel to and toward its surface. The fact that it is possible for light to be carried parallel to the black hole's surface does not, however, mean that light near its surface will necessarily orbit it.

As our earlier example suggested, imagine that the surface

of the Earth is smooth and completely covered with water. Now imagine that gravity is eliminated at one point on the Earth, such as at one of the poles. The result would be that gravity would force all of the water around the planet and it would leave the planet at this pole. The difference between water and light in this analogy is that light cannot leave the instance of origo that carries it; instead, the light energy will be compressed.

Because there is no outward force coming from the black hole, there is no inward force acting upon any instance of origo from the direction of the black hole. This lack of inward force acting upon any instance of origo would be akin to the pole in our water analogy, at which gravity was eliminated. Therefore, the light energy would be forced in the direction of the black hole, the path of least resistance.

Although the escape velocity is greater than the accepted value for the speed of light in a vacuum, insufficient velocity is not the reason that light cannot escape from a black hole; it is due only to the lack of origo moving away from the black hole, which would be necessary to carry it away.

The Ever-Changing Event Horizon

According to origo theory, origo carries light through the universe; light does not propel itself through the universe.

Without some condition to prevent it, origo is always in motion in straight lines, in every direction. For an event horizon to exist, it seems that origo would have to hover at a distance from the surface of the black hole.

How could it be possible for origo to hover at a distance from a black hole without falling in or escaping?

Origo is always in motion; therefore, it is not possible for it to hover at a distance from a black hole. Origo, which would normally tend to travel in a straight line that is moving parallel to the surface of a black hole, will curve toward the black hole because of the gravitational field surrounding the black hole. Origo that curves this way can take several different paths. The path of origo that passes close to the surface of the black hole will be curved so much that it will fall into the black hole. The path of origo that passes at a great distance will be curved, but not enough to cause it to fall into the black hole, and it will continue onward through the universe. Somewhere in between there is a place where the path of origo will be curved just enough to cause it to orbit the black hole without falling into it or escaping from it. At this radius is what we know as an event horizon.

Any light energy that is carried by the origo trapped in the event horizon will be trapped there also. This light energy cannot originate at the black hole; it must come from an outside source. Any light that is generated closer to the black hole than the event horizon will not travel outward, because there is no origo to carry it outward. Instead, it will fall into the black hole.

As matter and energy enter and pass through the event horizon, changes to the event horizon take place.

As energy and matter fall into the black hole, the mass of the black hole increases, which subsequently influences its gravitational field. Because of this, the radius at which the event horizon exists becomes larger. As this radius becomes larger, any light that is trapped within the old event horizon becomes light that is between the new event horizon and the black hole. When this occurs, the light is forced toward the black hole and ceases to exist as part of the event horizon. The event horizon is an ever-changing thing; its radius is continually increasing.

Matter at the Speed of Light

To help show what happens to matter near the surface of a black hole, let us consider a rock falling toward a black hole. The size and density of the rock is irrelevant. As it gets closer to the black hole, the gravity between the rock and the black hole becomes stronger. As the gravity becomes stronger, the rock travels faster. As the rock gets closer to the black hole, the inward force acting upon it, from the direction of the black hole, decreases.

Remember, according to this theory, the inward force holds every atom together by holding the energy, of which they are comprised, in a compressed state.

When the rock reaches a point where the inward force coming from the direction of the black hole is no longer strong enough to hold the rock's atoms together, all of the energy forming the atoms that make up the rock will be released and move in the direction of the black hole. At this point, the rock will become part of the electromagnetic spectrum and accelerate to and travel at the local speed of light until it reaches the surface of the black hole. The speed of light here is greater than 186,000 miles per second because of the intensity of the black hole's gravitational field.

The Light-Speed Limit

The speed of light will vary throughout the universe. If light were the only aspect of the universe it would travel at a single speed everywhere. Light, however, is not the only aspect of the universe, and its speed is affected by every other aspect of the universe. When light passes into a transparent medium that is denser than the medium that it was in, it slows down. When light passes from a denser medium into a less dense medium, it speeds up.

Light that is near to and traveling toward a gravitational field will travel faster than light that is far from and traveling toward the same gravitational field. Light that is near to and traveling away from a gravitational field will travel slower than light that is far from and traveling away from the same gravitational field. This is due to the reduction of origo coming from a gravitational field.

Light itself may reach velocities exceeding 186,000 miles per second when influenced by a gravitational field, but matter under all naturally occurring circumstances can never reach the speed of light, as it will become part of the electromagnetic spectrum before this occurs.

Matter above the Speed of Light

Is it possible for matter to reach or exceed the speed of light?

No matter what scientific theories one tends to believe, most would undoubtedly agree that as long as the conditions required for super luminal travel do not violate any of the laws of physics, it may be possible.

The theory of origo states that origo is infinite motion that moves at a finite velocity because of resistance due to origo in opposing directions.

If origo could somehow be diverted around a craft, thereby creating a sort of cosmic low-pressure area in front of the craft, it may be possible for macroscopic objects to exceed the speed of light. The keys to this happening are in understanding why the speed of light is the value that it is, and in understanding how to manipulate origo.

All current theories prohibit super luminal travel from happening, but none of these theories are anywhere near complete, nor do they explain why light travels through the universe. Moreover, none of these theories can explain the why of inertia or gravity. Therefore, with modern physics in such an incomplete state, it is much too early to definitively rule out this super luminal travel.

DID YOU KNOW?

There is a black hole at the center of our galaxy!

CHAPTER FIVE HIGHLIGHTS

1. Origo is moving toward black holes from every direction but is not moving directly away from them.

2. The reason that light cannot escape from a black hole is because of the lack of origo moving away from the black hole, which is necessary to carry it away.

3. The speed of light near a black hole will exceed 186,000 miles per second.

4. The speed of light varies throughout the universe.

5. Modern science is in such an incomplete state that it is much too early to rule out super luminal travel.

TRY THIS!

Keep on the lookout for the moon during the daytime. When you notice it, put on some dark sunglasses and, without looking directly at the sun, try to see the Earth, moon, and a glimpse of the sun all at the same time. As you do this, try to imagine what kind of views you might see on a distant world.

CHAPTER SIX

AN ATOMIC EXPLANATION

In this chapter

- We will briefly examine the currently accepted atomic model

- Be introduced to the optional laws of physics

- Consider the gravitational fields surrounding atomic nuclei

- Be introduced to transient energy and the electrosphere

- Consider the electron

- Consider origo's motion through matter

The Current Atomic Model

The currently accepted model of the atom consists of a nucleus, an electron cloud, and empty space. In this model, almost all of the atom's mass is contained within the positively charged nucleus; the remaining mass makes up the negatively charged electrons. Most of the atom's volume is the empty space between the electrons and the nucleus. The electrons exist within an electron cloud, which is not actually a cloud but the empty space surrounding the nucleus, which symbolizes the probabilities for the electrons' locations as they orbit the nucleus.

Something Here Is Not Quite Right

If there are electrons in orbit around the nucleus and only empty space between them, the atom could not exist with other atoms. Let us consider a hydrogen atom, which is purported to have only one electron in orbit in the empty space surrounding the nucleus.

The empty space between the electron and the nucleus of a hydrogen atom must be just that, empty space, for there would be no room for air or anything else to exist there. The empty space must be a complete vacuum; if it were not a complete vacuum, there would be resistance to the electron's motion, and its orbit would decay, causing it to collide with the nucleus.

Let us now imagine two hydrogen atoms next to each other. There are two separate positively charged nuclei with nothing but empty space between them, each having one orbiting, negatively charged electron. Now imagine that we somehow force these two atoms toward each other. If there is only empty space between them, then there is nothing to prevent both atoms from merging. One may say that the positive charge on the protons in the nuclei will repel each other, thus preventing them from merging. However, the electrons of each would surely collide with the protons of the other because they would attract each other, thus making the nuclei of both atoms electrically neutral. At this stage, there would be nothing to prevent the nuclei from colliding; instead of two hydrogen atoms, there would remain only one electrically neutral mass, consisting of most of the mass of both hydrogen atoms. Not only does this model not make sense, but it also violates basic laws of physics: Maxwell's laws of electromagnetism.

The Optional Laws of Physics

According to Maxwell's laws of electromagnetism, an accelerated charge will produce an electromagnetic wave and radiate energy away. Within the accepted atomic model, electrons are charged particles, and these electrons are in circular or semicircular orbits around the atomic nucleus. If these electrons are in orbit and they are charged particles, then they are accelerated charges and, therefore, must radiate energy in the form of an electromagnetic wave. Of course, if this actually happened, the electron would not exist for very long and, therefore, neither would the atom.

In 1913, the Danish physicist Niels Bohr, while working on the problem of the line spectra of gases, made an assumption that had a significant impact on the way the universe is viewed. This assumption gave him the solution to the problem, but it violated James Clerk Maxwell's laws of electromagnetism. Niels Bohr assumed that electrons could travel in circular paths without radiating their energy away in the form of electromagnetic waves. Because this approach allowed for the solution to the line spectra problem, the fact that it violated certain laws of physics was overlooked and it was accepted, though now we call them electron clouds instead of orbits. It is still accepted today, and much of what is believed about Quantum Theory is based upon Niels Bohr's assumption and its subsequent acceptance.

Physical laws are not valid only when we want them to be or when they fit our needs; they are either always valid or they are not laws at all. The laws of physics are not optional.

Consider this: is it more likely that the accepted model of the atom is correct and that these laws of physics do not apply in this one instance, or is it more likely that the accepted model of the atom is incorrect and that these laws of physics apply in all instances?

Understanding that the current model of the atom is just that, "a model," we can tolerate the fact that it violates known laws; however, we must not be content with it as it is. We must correctly describe the atom at some point.

The Nucleus' Gravitational Field

The following is a new description of the atom based upon the theory of origo. We know from observations and measurements that gravity is a significant force only when one of the objects is very large, such as a planet, so how could very small objects like atomic nuclei have extreme gravitational fields?

Objects that we see and touch every day are made of trillions of atoms, but these objects really do not have much mass in comparison to the moon or Earth. The nuclei within the atoms of these objects, however, are extremely dense.

As in the current atomic model, the nucleus is where almost all of the atom's mass is concentrated; it is extremely dense and consists of subatomic particles. The quantity and arrangement of these particles within the nucleus determine the properties of each atom. The nuclear material is so dense that a single teaspoonful would weigh about a billion tons on Earth. The extremely dense nucleus of an atom is comparable to a miniature neutron star.

Origo loses a large amount of energy to the subatomic particles that form atomic nuclei as it holds them in their compressed state, thus generating extreme gravitational fields. The nucleus is so small that this extreme gravitational field is extreme only at the atomic level, and an individual field is not noticeable on the scale of human senses. If the nucleus has an extreme gravitational field, would not an object with trillions of atomic nuclei, such as a person, also have an extreme gravitational field? Would not a pen with which we write stick so tightly to our hand that we could not remove it?

The gravitational fields of atomic nuclei are the same as

that of a planet, in the sense that the further one goes from the nucleus, the less influence the gravitational field has. The nuclei in the atoms of the pen, however, are so small and so far from the nuclei in the atoms of our hands that the gravitational fields of each are unnoticeable by us. Despite our inability to perceive these individual gravitational fields, they are still there; when many atoms are brought together in great quantities, they become very noticeable.

Transient Energy and the Electrosphere

If the gravitational field surrounding each nucleus is so extreme, what keeps the nuclei of two atoms apart?

We will now be introduced to a special kind of energy called transient energy; this is energy that does not exist in a semi-permanent state, as does the energy that forms atomic nuclei. This transient energy can be more easily changed from one form to another. Among others, kinetic energy, heat, electricity, and all parts of the electromagnetic spectrum are transient energy.

This transient energy surrounds every atomic nucleus. It is held in place by the extreme gravitational field surrounding the nucleus. It is similar to the atmosphere around a planet but much denser. Transient energy is compressed by the inward forces acting upon the atom. The mass of this transient energy, however, does not amount to much when compared to the mass of the nucleus that it surrounds.

We will refer to the transient energy surrounding an atomic nucleus as an electrosphere.

The gravitational field around an atomic nucleus is strongest at the surface of the nucleus and becomes weaker as the distance from the nucleus increases, just as it does for a planet, star, or any other body. Because of this, the electrosphere is denser near the surface of the nucleus than it is at the outer edge.

Since the compressed energy of the electrosphere is not nearly as dense as the nucleus, and since it becomes less dense farther from the nucleus, the atom's gravitational field does not increase substantially as the size of the electrosphere increases. This means that every electrosphere has a maximum size that it can be, and this size is determined by the mass of the nucleus.

Origo is traveling toward the atom from all directions (the inward force); this is what holds the nucleus together and compresses the transient energy around it. Origo is also traveling in all directions parallel to the surface of the nucleus. Origo traveling in these parallel directions will shear away any transient energy that is not bound very strongly by the gravitational field of the nucleus. Therefore, a single isolated atom will be nearly spherical, depending upon the shape of its nucleus.

Regardless of the size or type of atom, the gravitational force required to prevent the electrosphere's transient energy from being sheared away by origo moving in directions parallel to the atom's surface, will be the same. Therefore, the density of the transient energy at the outer edge of every atoms electrosphere will be the same.

We can now answer the question: "If the gravitational field of each nucleus is so extreme, what keeps the nuclei of two atoms apart?"

It is the atoms' electrospheres that keep them apart. In

order for two or more nuclei to reach each other, some or all of their electrospheres would first have to be displaced. If a small amount of this compressed energy were removed from one or both atoms, it would be possible for the nuclei of the two atoms to get a little bit closer together, thus forming a gravitational bond between them. Likewise, if enough transient energy is added to atoms that have formed a bond, their electrospheres will expand, thus increasing the distance between their nuclei and making it easier to separate them.

Together, a nucleus and an electrosphere form an atom. When atoms have a cohesive bond, the nuclei of the atoms share electrospheres. In this arrangement, the total compressed energy in the combined electrospheres of two atoms is less than the total compressed energy in the electrospheres of two individual atoms.

A quantity of lead atoms at room temperature is a solid but soft metal, very malleable, but solid, nevertheless. When transient energy such as heat is added to this lead, it becomes even softer until it changes to a liquid state. What happens as heat is added to the solid lead is this: the quantity of transient energy surrounding the nucleus of each lead atom increases, causing each atom's electrosphere to expand. This expansion causes the distance between nuclei to increase, thereby diminishing the gravitational bond between them and enabling them to move about more freely.

This would be in line with the idea that the atoms of gases do not have a substantial gravitational bond between their nuclei. Liquids, however, have a small gravitational bond between the nuclei of their atoms, and solids have a substantial gravitational bond between the nuclei of their atoms.

If the solid, room temperature lead were cooled instead of heated, it would become harder and more solid; as the lead is cooled, transient energy is removed from the lead atoms' electrospheres, causing their nuclei to move closer together and thus increasing the gravitational bond between them. This is the reason that most objects become flexible when they are heated and rigid when cooled.

This does not mean that the atoms' electrospheres could be removed if one were to keep cooling the lead, leaving only the nuclei.

At a temperature of absolute zero, atomic electrospheres are as large as they can possibly be without energy being supplied to them from an outside source, and as small as they can possibly be without energy being physically removed from them by an outside source.

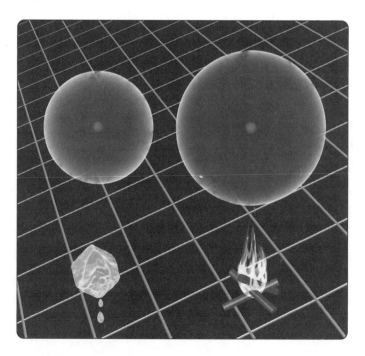

When transient energy is added to an atom, its electrosphere expands, but its nucleus does not.

Let us consider the electrospheres of the atoms of a gas. If a gas is heated to a temperature far above absolute zero, it is the transient energy (heat) expanding the gas atoms' electrospheres that causes the nuclei of the gas to spread so far apart that the gravitational bond between them is virtually nonexistent. When the gas is cooled, some of the transient energy is removed from the atoms' electrospheres, thus causing their nuclei to reform the bonds between them; this is why a gas will liquefy when it is sufficiently cooled.

No Electrons Inside

If every atom has an electrosphere, then there is no place for an electron to exist within it. How can we reconcile this with the fact that electrons have been detected and shown to exist?

The fact that electrons have been shown to exist does not necessarily mean that they exist within atoms. Every experiment that produces electrons produces them on the outside of atoms, never within. This is because electrons do not exist within atoms.

As was stated in the previous section, origo "will shear away any transient energy that is not bound very strongly by the gravitational field of the nucleus." Therefore, the outer edge of an atom's electrosphere has a definite boundary, within which all of the energy is compressed. This means that every individual atom's electrosphere has surface tension.

Let us look at light colliding with an atom and the effect

that it has on the atom. As light collides with an atom, it first strikes the surface of the atom's electrosphere. If the light has enough momentum to penetrate the surface tension, the light energy will enter the electrosphere. If the light does not have enough momentum to penetrate the electrosphere's surface tension, it will bounce off. When white light strikes an atom, some of its energy will usually penetrate the electrosphere and some of it will be reflected.

As light energy penetrates the electrosphere, the electrosphere tends to expand. This extra energy cannot exist within the atom indefinitely. The gravitational field of the nucleus allows for the compression of a specific quantity of energy within the electrosphere. This extra transient energy will eventually be sheared away by origo.

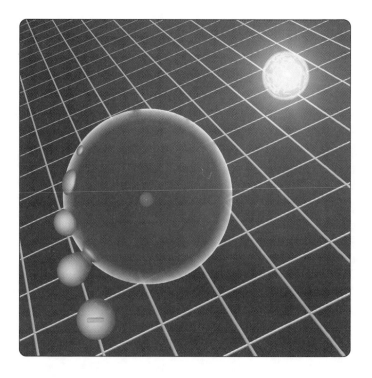

If transient energy is applied to an atom under the proper conditions, the extra energy will leave the atom in lumps. It will leave the electrosphere at the point of least resistance. At this point, the extra energy can be forced out of the electrosphere without losing density or its surface tension; it is held in its compressed state by the inward force acting upon it. It leaves in the form of an energy bubble of the same density as the outer edge of the electrosphere, growing out of the side of the atom's electrosphere.

This bubble of compressed energy is what we know as an electron. It is a particle of transient energy that has no nucleus and exists only outside of atoms. It has surface tension and can easily be caused to burst, thereby releasing its energy.

How is it that electrons move freely through metals?

Electrons do not exist within atoms; therefore, they do not move freely through metals. The atoms of a metal share their electrospheres with all adjacent atoms, thus forming a cohesive bond through which transient energy can flow.

When energy is added to the atoms of a metal, their electrospheres realize an energy increase and expand. This energy follows the path of least resistance to the surface, and electrons can be generated as the energy leaves the surface of the electrospheres.

When an electron is destroyed and the compressed transient energy is released, it can take any of several forms, depending on how it is released. It can be released to origo as part of the electromagnetic spectrum, which means that under certain conditions the energy contained within an electron can be released in such a way as to produce an interference pattern.

Origo's Motion through Matter

Origo is the force that carries light through the universe, and it is slowed down by passage through matter, as evidenced by light slowing down as it enters a denser medium. We can see that origo slows down as it passes through matter, and now we will reason why it does.

We have reasoned that origo is diverted by areas of density greater than its own normal density; therefore, it makes sense that origo would also be diverted by matter, since matter is made of origo that has been compressed to varying and extreme densities.

The density of an atom's nucleus is so great that it is very difficult for origo to travel through it. If origo cannot easily travel through the nucleus, it must take a while to get through. Therefore, some origo will go around it, causing origo to travel a greater distance between two points than it would if it traveled in a straight-line path.

If origo is traveling around the atom's nucleus, it must be traveling through the atom's electrosphere. The electrosphere is compressed energy that origo can travel through, but not as easily as it can travel through a "vacuum." Therefore, this also adds to the slowing down of origo as it travels through matter.

As origo moves through matter, it must flow around the atoms' nuclei, much like a river flowing around rocks. Origo flowing around the nuclei will take the path of least resistance, which will be the furthest or halfway point between nuclei, where the electrospheres are the least dense.

This description of the atom is by no means complete. The way in which origo moves within and around atoms is more complex than is described within this text. More research and theorizing will need to be done in order to put a complete atomic picture together, but as this field is explored further, it should help to make clearer the exact mechanism for electricity, magnetic fields, and superconductivity, to name just a few problems still demanding solutions.

> ### DID YOU KNOW?
>
> One teaspoonful of material from a neutron star would weigh about a billion tons on Earth!
>
> The numbers are large and difficult to comprehend. Therefore, I will help you put it in perspective.
>
> A single Nimitz class aircraft carrier displaces 97,000 tons. If we could take 10,309 of these aircraft carriers and crush them down until they all fit into a single teaspoon, they would reach the density of a neutron star!

CHAPTER SIX HIGHLIGHTS

1. If the currently accepted atomic model were correct, two or more hydrogen atoms could not exist together.

2. Atomic nuclei are held together by gravity.

3. Atomic nuclei are surrounded by compressed transient energy, which forms an electrosphere.

4. The outer edges of all atoms' electrospheres are the same density.

5. Materials expand when heated because, as energy is added to them, their electrospheres expand, increasing the distance between nuclei.

6. Electrons do not exist within atoms but can exist outside of them.

7. Origo moves between nuclei much like river water moves around rocks.

TRY THIS!

Put a ceramic coffee cup full of cold water in the microwave. Before you close the door, identify a mark on the cup that you will be able to find again easily, and note its elevation relative to the water level. Now close the door and heat the water to the temperature of a hot cup of coffee. Find the mark on the cup again and notice how much the water molecules have expanded.

CHAPTER SEVEN

THE MAGNIFICENT MAGNETIC FIELD

In this chapter

- We will consider where the energy within a magnetic field originates

- How magnetic fields are generated

- Why like poles "repel" each other

- Why opposite poles "attract" each other

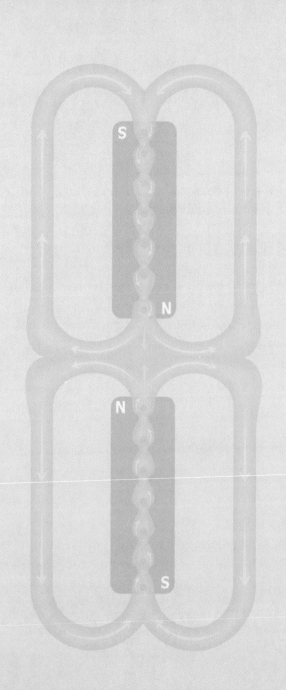

The Energy Within a Magnetic Field

Magnets are very different from most matter around us. They seem to contain their own energy source, allowing them to attract things such as iron, electrical fields, and other magnetic fields. At the same time, they also repel things such as electrical fields and other magnetic fields.

You can place a magnet on the underside of a piece of iron and it will hang there indefinitely, seeming to defy gravity. Although the magnet may defy gravity, it does not defy the laws of physics. It seems that the magnet should fall, but there is something holding it up that follows the laws of physics.

According to this theory, origo is all the energy of the universe; therefore, origo must be the force keeping the magnet suspended on the underside of the piece of iron.

Magnetic fields occur in conjunction with matter; they do not occur as isolated fields without matter present. Therefore, we can reason that magnetic fields occur somehow, due to the presence of matter.

All of the energy of the universe exists in two states: origo (the inward and outward forces) and compressed origo (transient energy such as heat and light and semi-permanent forms such as subatomic particles and atomic nuclei).

Because magnetic fields occur only in conjunction with matter, and because matter consists of (semi-permanent) atomic nuclei and (transient) electrospheres, and because origo acts upon all matter to hold it in its compressed state, we can reason that somehow all forms of origo are involved in the existence of magnetic fields. We need only identify the role that each plays in their existence. Magnetic fields are fields that are continuously in motion. If the continual motion of a magnetic field was caused by the energy of which the magnet's atoms are comprised, the magnet would eventually deplete its energy and its field would cease to exist. This does not happen.

Because the energy within a magnetic field cycles indefinitely without a reduction in the energy of the magnet's atoms, we can reason that the energy of the magnet's atoms is not responsible for the motion of the energy within a magnetic field.

The energy of the magnet's atoms has been ruled out as being the force behind the continuous motion of a magnetic field. Therefore, we are left with only one other option, origo.

According to origo theory, gravitational fields are "fields of origo." They are due to an imbalance between the inward force and the outward force in the region surrounding a body.

If magnetic fields consist of origo, and a wire moved through a magnetic field produces electricity, then a wire moved through any field of origo should produce electricity. Therefore, if the statement "a wire moved through any field of origo should produce electricity" is correct, then a wire moved through a gravitational field should produce electricity. We know that this is not the case. A wire moved through a gravitational field does not necessarily produce electricity. Therefore, we must conclude that the above statement is incorrect.

We have reasoned that origo is the force that is behind the continuous motion of a magnetic field and that a wire moved through a field of origo does not necessarily produce electricity. We know, however, that a wire moved through a magnetic field does indeed produce electricity. Therefore, we can reason that a magnetic field consists of more than only origo.

Atoms are not cycling around within magnetic fields; therefore, we can rule out (semi-permanent) atomic nuclei as being the other factor of which magnetic fields consist. Thus, we are left with only one other option, transient energy.

Our reasoning so far has led us to conclude that magnetic fields consist of transient energy that is put in motion by origo.

How Magnetic Fields Are Generated

A magnet is an energy structure that is unique, in that, due to the properties of its atoms, origo passes through it more freely from one direction than it does from the opposite.

The inward force is moving toward the magnet's atoms equally from all directions (unless, of course, it is in a gravitational field). For some reason, the atoms require more origo from one direction than from the opposite to hold them in their compressed state. Because of this, the outward force from each atom is less in one direction than in the opposite.

Because the outward force is less in one direction than in the opposite, the effect of the inward force on the transient energy of the electrospheres of the magnet's atoms differs on opposite sides of the atoms.

The sides of the atoms with less outward force in effect have stronger gravitational fields than do the sides that have more outward force moving away from them. This difference in the strength of the gravitational field on either side of each atom results in a dynamic electrosphere.

The transient energy of the side of the electrosphere with the stronger gravitational field is forced to flow around the nucleus toward the side of the electrosphere that has the weaker gravitational field; this is the path of least resistance.

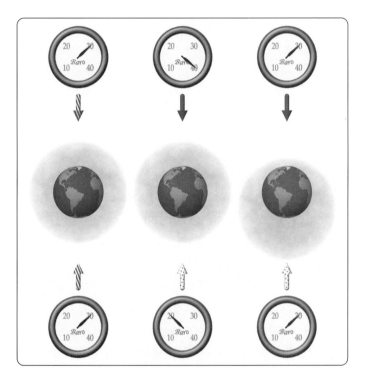

Imagine if Earth's gravitational field were stronger in one hemisphere than in the opposite. In a world such as this, the atmospheric pressure at sea level would be greater in one hemisphere than in the opposite. As a result, the atmosphere in the hemisphere with greater pressure would move around the planet and into the hemisphere of lower pressure. The air within the hemisphere of lower pressure would be displaced by the influx of air from the hemisphere of greater pressure and would have to go somewhere. The accumulation of air in the one hemisphere would become more than the planet's gravitational field could normally contain, and much of it would be forced out of the normal limits of the atmosphere.

Just as the air would be forced out of the normal limits of the atmosphere, so, also, the transient energy is forced out of the normal limits of the magnet's atoms' electrospheres.

It is likely that all or many atoms behave in the manner described above, and that only the atoms of certain materials can all be aligned in the same direction so that the net effect is a large magnetic field.

If the atoms of a material are aligned in various directions, the effects of the energy flowing between them may cancel each other out, and the net effect would be very small and possibly unnoticeable. However, if the atoms of a material can all be aligned in the same direction, the result would be a large magnetic field.

The problem that the magnet's atoms now face is that if their transient energy is being forced away from the atoms and ultimately out of the magnet itself (extra-atomic transient energy), the atoms will collapse unless there is a steady supply of transient energy coming in.

In a material whose atoms are aligned in various directions this is not a problem; the transient energy of each atom will simply follow the path of least resistance into the next atom without leaving the object that is comprised of these atoms.

The magnet's atoms, however, do not collapse, because there is a steady supply of transient energy to the side of the atoms with the stronger gravitational fields. This supply of transient energy is the same energy that is ejected from the

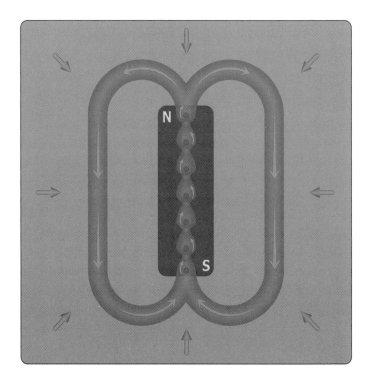

opposite side; it finds its way back around the magnet and into the other side.

As we concluded in the previous section, magnetic fields consist of transient energy that is put in motion by origo. We have so far identified the source of the energy within a magnetic field and how it is produced, and we also know that origo makes it move. Now we need to identify just how origo makes the transient energy move within a magnetic field.

Because, in a magnet, the outward force from each atom is less in one direction than in the opposite and because the atoms are all aligned in the same direction, the outward force from the magnet as a whole is less in one direction than in the opposite.

Because origo, the inward force, is moving toward the magnet's atoms equally from all directions and because the outward force from the magnet is less in one direction than in the opposite, there is an imbalance between the inward and outward forces on one side of the magnet, and the inward and outward forces on the opposite side. This imbalance causes the inward force to act upon the greater outward force by diverting it back around the magnet toward the area of lesser outward force; this is the path of least resistance.

The transient energy that is ejected from the magnet is caught in the current of the outward force as it is channeled back around the magnet, where it re-enters the magnet and cycles through again.

The side of the magnet where the inward force flows through more freely is the magnet's south pole. The side of the magnet where the inward force is retarded is the magnet's north pole.

The previous description explains why the field lines of a magnet enter at one pole and exit at the other and also why the field lines form closed-loops. This continuous cycle of transient energy through the magnet and the diversion of origo from its straight-line path are what create a magnetic field.

Repulsion between Two North Poles

Why do the north poles of two magnets "repel" each other?

As transient energy is ejected at the north pole of each magnet, origo forces it around the magnet, resulting in a "field" that extends some distance out from the magnet.

The fields of extra-atomic transient energy around each magnet are in equilibrium; although the normal flow of origo has been disrupted by the atomic structure of the magnet, the energy field flowing around and through the magnet has a certain motion, density, and size at which it normally exists when it is isolated from other objects. All of the forces involved are balanced in this normal state.

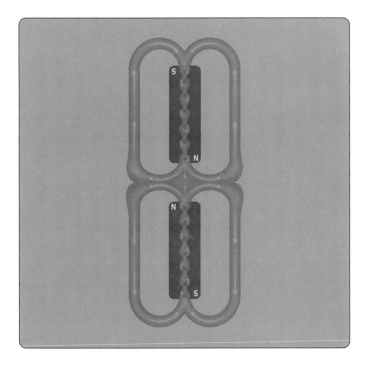

When the north poles of two magnets are forced together, the two fields are forced into a region that is smaller than the total size of both individual fields. This causes pressure between them because it results in an even greater density of extra-atomic transient energy and origo in this area.

Because the atoms of the magnets are the instruments causing the fields by altering the normal flow of origo, the pressure between them forces the magnets apart. A rough comparison of this effect is as follows: if two separate cubic feet of air molecules of equal pressure and temperature are forced together into a volume equal to one cubic foot, pressure will result. This pressure will push against whatever is holding the air in this compressed state: the walls of the container. We can consider the magnets to be "the walls of the container," and if the walls can move to relieve the pressure, they will.

Repulsion between Two South Poles

If the south pole of a magnet is the area where origo flows through more freely, why do the south poles of two magnets "repel" each other instead of "attracting" each other?

The transient energy that is ejected from the north pole must flow around the magnet and into the south pole, causing an increase in extra-atomic transient energy at the south pole as well as at the north. Because of this extra-atomic transient energy at the south poles of a magnet, when the south poles of two magnets are forced together, the effect is almost the same as when two north poles are forced together. The difference is that, for south poles, the repulsion may be slightly less than occurs at the north, due to the ability of origo to flow more freely here.

Attraction between a North and a South Pole

Why do the north pole of one magnet and the south pole of another magnet "attract" each other?

When the north pole of one magnet is placed next to the south pole of another, the path of least resistance for the extra-atomic transient energy that is flowing from the north pole, around the magnet, and into its own south pole, changes. Instead of flowing around the magnet that it came out of and back into its own south pole, it flows around both magnets, and into the south pole of the other magnet. These two magnets act as one, and the flow of origo and the transient energy continues as it would if there were only one continuous magnet.

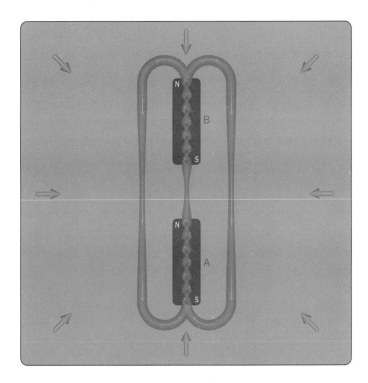

When origo and transient energy flow from the north pole of magnet A into the south pole of magnet B, they fill the space into which origo and transient energy from the north pole of magnet B would normally flow. This means that origo and transient energy from the north pole of magnet B cannot continue their normal flow into the south pole of magnet B. At this stage, origo and transient energy from the north pole of magnet A are no longer flowing around magnet A and into its own south pole. Therefore, the path of least resistance for origo and transient energy from the north pole of magnet B is around both magnets B and A, and into the south pole of magnet A. Because the inward forces acting upon both magnets are acting on them from all directions, their combined magnetic field cannot be any larger than necessary. Therefore, when the north pole of magnet A comes near the south pole of magnet B and the fields begin to merge, the two magnets are pushed together by the inward forces.

It is easier for origo and transient energy from the north pole of magnet A to flow into the south pole of magnet B than it is for them to flow around and into the south pole of magnet A. Therefore, when the merging of the two magnetic fields begins to take place, if there is space between the magnets, the merging fields in this area become less dense than the separate individual fields. The result is a sort of low-pressure area between the magnets, which allows the inward forces pressing on both magnetic fields to push the magnets together.

Magnetic Monopoles

Magnetic monopoles are something that scientists theorize about and search for. They are a single magnetic pole such as a north pole alone without an accompanying south pole, or vice versa.

According to origo theory, north and south poles are the result of an imbalance of the outward force between two sides of an object's atoms.

The opposing poles are the result of the inward force moving toward the atoms of the magnet equally from all directions, while the outward force from the magnet is less in one direction than in the opposite. This results in an imbalance between the inward and outward forces on one side of the magnet, and the inward and outward forces on the opposite side. This imbalance results in the inward force acting upon the greater outward force and diverting it back around the magnet toward the area of lesser outward force; this is the path of least resistance. For the greater outward force to do other than to move into the area of less lesser outward force and maintain the balance would be to follow the path of greatest resistance.

All energy of the universe and the universe itself consist of origo. Origo is a conservative force, and because of it, every object and force, every aspect of the universe, is always in equilibrium.

In order for a monopole to exist, origo, at some point, would simply have to cease to exist, thereby preventing it from equalizing and forming both north and south poles. This would violate the first law of thermodynamics. Therefore, search as we may, magnetic monopoles will never be discovered.

DID YOU KNOW?

When two magnets "repel" or "attract" each other, field propulsion is taking place!

CHAPTER SEVEN HIGHLIGHTS

1. Origo passes through magnets more freely from one direction than from the opposite.

2. Origo causes transient energy to cycle through magnets, resulting in magnetic fields.

3. Magnetic monopoles cannot exist in the universe and, therefore, will never be discovered.

TRY THIS!

Take nine disk-shaped magnets and arrange them in the form of a square. To do this you will need to alternate the polarity of each magnet. After it is assembled, hold the center magnet and rotate the rest. Notice how they all act just like gears without teeth. Remember, you have other things to do than play with these magnets all day!

CONCLUSION

When I was a young boy, I thought that every aspect of the universe was understood by scientists. I remember reading about Sir Isaac Newton and imagining the thrill that he must have had as he experienced and studied gravity. I remember thinking that these types of discoveries and realizations could not be made by me, because they had already been made by others. These thoughts saddened me.

Through this Logical Universe project, I see a whole new world about to open up for us all. I am excited about the many opportunities there will be for people in the research and development fields, as well as about the hands-on construction and operation of the infrastructure that will support the new technologies outlined in this book. There will be opportunities for everyday people to experience things that are now only the stuff of science fiction stories.

I believe that in my lifetime I will see this technology emerging and will be able to experience some of it. I expect, at the very least, to be able to go to the edge of space, where I can see Sol, Luna, Earth, and the stars, all at the same time.

If technology based upon the ideas presented in this book

is currently being researched by any government entities, it is undoubtedly classified and would not be made available to the general public for decades. Now, however, having these ideas in the hands of everyone who owns a copy of this book should speed up the process between conception and general availability of these technologies.

I hope this book has caused you, the reader, to realize that the universe is indeed a remarkable and astoundingly simple thing that anyone can understand and that you are not only in the universe, but of the universe. If this book has not caused that realization, read it again!

Each one of us who ponders the workings of the universe is, in effect, the universe considering itself.

Michael F. Jones

ACKNOWLEDGEMENTS

Developing the concepts described within the pages of this book was a labor of love for me, but it was very trying for my wonderful wife, Sonia, and for our two boys, Abraham and Alexander, as it consumed most of my available time for several years. Without your patience, understanding, and support, *Logical Universe* would still be a work in progress. Thank you, Sonia, Abraham, and Alexander!

I am very grateful for, and would be remiss if I failed to mention, this great nation that I am honored to call my home. The peace, security, freedom, and abundant resources that the United States of America provides us should never be taken for granted! Without these wonderful gifts, many of the comforts and conveniences that we Americans have grown up with and have often taken for granted would not be possible, and the writing of this book would have been a very difficult thing. I am indebted to all who help keep this country great!

I am grateful for my brothers and sisters—Cynthia, Earl, Tamara, Joann, and James (who may someday see the light)—who provided support and many hours of stimulating conversation regarding the universe.

I am thankful that Barron's Educational Series, Inc., allowed the use of excerpts from one of my favorite books:

Physics the Easy Way, Second Edition;
Robert L. Lehrman
1990, Barron's Educational Series, Inc.
ISBN: 0-8120-4390-1

Thanks to Bantam Books for allowing the use of excerpts from a very interesting and mentally stimulating book:

The Universe in a Nutshell, Stephen Hawking
2001, Bantam Books
ISBN: 0-553-80202-X

Thanks to NASA and to all others who provide websites with valuable information that is freely shared with the world.

Without the host of people who have contributed to the knowledge base of the world, this book could not have been written. Some of them are recognized below, and for all scientists that I fail to specifically mention, I now sincerely thank you for all of your contributions.

Nicolaus Copernicus,
Your scientific contributions propelled mankind forward and helped to curb our collective arrogance. You showed us that the solar system does not revolve around the Earth, and that the Earth is just another planet that is revolving around the sun. I admire you for the great courage you displayed as you went head-on against the prevailing powers of the church,

who ignorantly and blindly believed that the Earth was the center of the universe.

Galileo Galilei,
One of the very first things that I can remember as a young boy that really opened up my mind to science is the story of you dropping balls of different size and weight from the top of the Leaning Tower of Pisa. Although you are well known for many other major contributions to science, the leaning tower experiment is how I connect with you. I also admire you, just as I do Nicolaus Copernicus, for standing up for truth and real understanding in the face of great adversity.

Sir Isaac Newton,
As a young man you were sitting in church, watching a chandelier sway gently in the breeze that was coming through a window. You were curious about why the chandelier moved the way that it did. This simple observation sparked your imagination and interest in the laws of motion. At least that is the way it was conveyed to me as a young boy.

Your dedication to scientific study could have come from nothing less than a sincere love for the universe, something that both you and I share.

The laws of motion that you defined had far greater influence on the world than any of your contemporaries could have imagined. I most appreciate your illumination of the fact that gravitational mass and inertial mass are equal.

Michael Faraday,
As a young boy, I read your biography and met a man who did

what he loved and loved what he did. You did not let poverty hinder you. You just did what you had to do, but, truly, it was something electrifying!

Without your groundbreaking work, I probably would not be typing on this laptop computer right now.

Your biography and your work with electricity so fascinated me as a young boy that I am sure it was a major factor in cultivating my love for the universe. My book *Logical Universe* describes physical reality by the same field that you worked with long ago.

Albert Michelson and Edward Morley,
Your brilliant experiment that we all know as the "Michelson-Morley experiment" is a landmark in modern physics. Although the results of the experiment have been misinterpreted for many years, it is nevertheless an extremely valuable experiment. Because the split "beam of light" did not produce an interference pattern when it was brought back together, we learned that light is not a wave.

Albert Einstein,
You are one of the most notable physicists of all time. Your brilliant, world-changing work stands alone. Your search for a comprehensive field theory that explains physical reality, a search that drew much criticism and little support, is something that will always stand as a testament to your remarkable intellect. When nearly the entire misguided world followed the illogical path of Quantum Theory, you stood your ground. You knew that the universe is logical and were not swayed. You also showed us that matter and energy are, in fact, the

ACKNOWLEDGEMENTS

same thing. You truly were a great man!

Edwin Hubble,
Your name is synonymous with clarity. You opened our eyes to our place in the universe when you showed us that galaxies exist all around us. You and Vesto Slipher also determined that the light from every galaxy is either blue- or red-shifted. This observation will help us understand the nature of light. Now, the most powerful telescope ever built by man is named after you. With the Hubble Space Telescope we see fantastic images of our universe that are so clear and awesome that they are almost unbelievable.

Max Plank, Ernest Rutherford, Niels Bohr, Erwin Schrödinger, Werner Heisenberg, Wolfgang Pauli, Paul Dirac, Tomonaga Shin' Ichiro, Richard Feynman, Julian Schwinger, Stephen Hawking, and all others who contribute to Quantum Theory, Much scientific advancement has been made as a result of your work. These advances have paved the way for many new technologies. The statistical mathematics of quantum physics, although very helpful in making predictions and calculations about what is happening, does not supply us with an accurate picture of the universe. Without your efforts, however, this vital type of statistical mathematics might never have been developed, scientists might not have been sidetracked for so long, and *Logical Universe* might have been written long before I arrived on the scene. You have saved this work for me.

Thank you all!

PREDICO

In 2006, just over one hundred years after Albert Einstein unveiled the special Theory of Relativity, Michael Jones published *Logical Universe*. The book introduced origo theory and showed the human race how simple the universe really is. Origo theory revealed the elusive mechanisms of magnetic fields, which gave research scientists a platform from which to unlock the secrets of superconductivity. This resulted in the Giant Electrical Component Corporation (G.E.C.C.) developing the first economical high temperature superconductive wire in the spring of 2007. This wire enabled electricity transmission to take place over long distances without resistance. It remained superconductive up to 93 degrees Fahrenheit. This "93 wire," as it was called, was a vast improvement over the old copper wire that had been used for more than a century, but it had its limitations. In deserts, perpetually hot regions, and whenever a hot spell struck more temperate areas, the superconductive property of 93 wire was lost, which caused rapid overheating and meltdown of the wires at times. Therefore, for much of the planet's electrical transmission needs, 93 wire was not a feasible alternative to copper.

Although 93 wire found countless uses in many applications, it was not suitable for widespread power transmission. In 2009 a relatively unknown wire manufacturer, Copper Extrusion Corporation, introduced the Holy Grail, a wire they called "Thermal Conversion Wire." This new wire would convert heat into electricity much like a thermocouple, and cold temperatures had no adverse effects on its superconductive property. The instant the wire began to realize an increase in temperature, the heat was converted and additional electricity was added to the system. This was very helpful in hot regions.

Within eighteen months after introducing thermal conversion wire to the world, the Copper Extrusion Corporation had purchased seven major competitors, licensed its manufacture to numerous other wire makers, and become the world's largest, most productive wire maker and the fastest growing company anywhere.

Thermal conversion wire enabled the transmission of electricity over vast distances with zero loss due to resistance. This sparked the construction of immense solar power plants in desert regions all over the globe. These vast, barren regions quickly became the oil fields of the future as the energy produced there began to supply cheap, clean, unlimited energy to power the new generation of electric automobiles as internal combustion machines began to be phased out.

Thermal conversion wire was also directly responsible for the 2011 Planetary Power Supplement Act PPSA, a global effort to supply electrical power to the most needy places on Earth. This simple effort effectively increased world peace substantially, as it gave billions a renewed sense of hope and a feeling that they were valued by the rest of the world. And those that had already been living in developed nations felt the goodwill as well, and more seeds of peace than ever before were sown. Thanks to thermal conversion wire and the PPSA, the old adage "it is better to give than to receive," is no longer a half-believed maxim but is practiced by the majority.

Also in 2011, the innovative Scaled Composites team, headed by Burt Rutan, that had won the "X Prize" seven years earlier made the very first test flight of a revolutionary field propulsion system that was described in some detail in *Logical Universe*. Their team was eleven months ahead of NASA's first successful test flight. NASA's field propulsion program had gotten off to a slow start due to much skepticism in the scientific community (especially from quantum physicists) and the reluctance to reduce or cut other more popular programs.

NASA's late arrival on the scene resulted in its reliance on the X-Prize team for propulsion technology, as the X-Prize team had obtained utility and design patents first.

Along with their steadily improving field propulsion engines, it is rumored that the X-Prize team is once again outpacing NASA as it prepares to test its Red Shift Communication System early next summer. If the tests are successful, real-time communication will become possible for the faster-than-light manned craft that are on the drawing board.

Field propulsion made liftoff and re-entry into Earth and other planets' atmospheres so safe that within two years of the first field propelled flight, a first-class luxury space tourism company was open for business with cruises to the moon, where the first Hilton (polymer dome constructs with all the creature comforts, including regular other-worldly concerts by artists such as Ronnie James Dio, Celine Dion, and Alabama) was established in Bailly Crater.

Now with cruises to as far away as Pluto taking place regularly, planetologists swarming every celestial body in Earth's neighborhood, new resorts springing up all over the solar system, and human exploration venturing to nearby star systems, it is just a matter of time until we hear the news of a great long-range, manned interstellar mission.

Hope, opportunity, and adventure are everywhere. What a fantastic time to be alive!

TAKE THE GRAVITATIONAL CHALLENGE

The Gravitational Challenge is aimed toward all physicists, teachers, authors, institutions, and individuals who claim that gravity is due to objects exerting a pull upon one another.

The purpose of the gravitational challenge is to bring to light the fact that modern science does indeed make ungrounded assumptions and then subsequently portrays them as fact. It is not intended to simply criticize modern science; it is designed to help move modern science from the rut it is stuck in and on to a more productive track. This practice of making ungrounded assumptions and portraying them as factual is an insult to all people who trust in modern science, many of whom unquestionably believe what their teachers and the scientific community tell them.

All who claim that gravity is due to objects exerting a pull upon one another are hereby being called out!

You are now challenged to either demonstrate this ungrounded claim by experiment or observation or to cease from making it!

One experiment that may be presented as evidence to support this groundless claim is the Cavendish experiment. This is an experiment that was performed by Henry Cavendish in 1798, in which he used lead balls and a torsion balance to determine the value of *G*. This experiment gave us valuable information however, it does not demonstrate that objects exert a pull upon one another any more than a falling object demonstrates it.

The fact that gravity exists is not being challenged; only the often-invoked statement that objects exert a pull upon one another is being challenged. Many would dismiss this challenge as unnecessary, but if we truly want to understand gravity and our universe, we must be honest with ourselves and be vigilant, lest we are blinded by our ignorant assumptions.

The Cavendish experiment clearly demonstrates that objects, for some reason, tend to move toward one another. But there is no component of the experiment, other than the interpreter's assumptions, that indicates that the objects are pulling on each other. Also, there is no component of this or any other experiment that excludes the possibility of objects tending to move toward each other because of an outside force pushing them toward each other.

In addition to the challenge offered to all who claim that gravity is due to objects exerting a pull upon one another, **all students and recipients of this ungrounded claim are now challenged to uphold your self respect as free thinkers** and to implore the providers of this ungrounded information to cite **a single** experiment or observation that supports it!

If you would like to contact Inertial Press with stories of your gravitational challenge efforts, they will be welcomed. All stories submitted should follow the guidelines listed in "Contact the Author."

GLOSSARY

The definitions listed here are not necessarily the only definitions available for the listed words and terms, but are the definitions intended to be used within this book.

Absolute Zero
0 Kelvin; The lowest temperature theoretically possible.

Accelerated Charge
An electrical charge that is accelerated.

Acceleration
A change in magnitude of velocity or direction.

All-Pervading Ether
A theoretical substance that, in the early twentieth century, was believed to exist throughout the universe at absolute rest.

Andromeda Galaxy
The Milky Way's closest neighbor, also known as M31.

Artificial Gravitational Field
A gravitational field produced by electrical and mechanical means rather than by a massive body.

Atmospheric Pressure
Pressure which results from the weight of a planets atmosphere.

Atomic Model
A representation of how atoms are believed to be.

Attractive Force
A force that pulls.

Balance
Equality between forces or objects.

Big Bang
A theoretical event whereby the universe came into existence.

Binary System
A system consisting of two stars revolving around each other.

Black Hole
The densest objects in the universe. Energy structures so dense

that origo loses all of its energy holding them in their compressed states.

Blue-Shift
A phenomenon in which light is shifted toward the blue end of the electromagnetic spectrum.

Bursters
Massive gamma ray explosions; the most powerful explosions in the known universe.

Charged Particles
Particles such as electrons that have an electrical charge.

Closed Loop
A condition whereby all events are the cause and result of all other events; a condition that has no beginning or end.

Completely Fluid
Readily changing, not fixed, flowable on any scale.

Corpuscular Nature of Light
The granular nature of light; consisting of particles.

Cosmic
Relating to the size of the universe.

Cosmological Expansion
The expansion of the fabric of the universe.

Cosmology
The study of the universe.

Curvature of Space-time
The bending of the fabric of the universe.

Deceleration
Acceleration that results in a lower velocity.

Deformational Energy
Energy stored in an object by distortion of the object, as in a spring.

Density
The amount of matter or energy within a specific volume.

Diffraction Grating
A device consisting of a transparent material with many fine slits, which is used to analyze light.

Doppler Effect
The change in the frequency of a sound or other wave caused by movement of the source relative to the observer.

Doppler Shifts
The shifting of the frequency of a wave as described by the Doppler Effect.

GLOSSARY

Double Slit Experiment
An experiment wherein light is analyzed by passing it through a screen with two slits.

Electromagnetic Energy
Energy related to electricity and magnetism.

Electromagnetic Field Generator
A device that generates electromagnetic energy.

Electromagnetic Fields
An extent of space wherein electromagnetic energy is present.

Electromagnetic Spectrum
The full range of frequencies of electromagnetic energy.

Electromagnetic Wave
A wave consisting of electromagnetic energy.

Electromagnetically Induced Imbalance
An imbalance caused by electromagnetism.

Electron
A negatively charged particle said to exist within atoms.

Electron Cloud
A region representing the probabilities of the locations of electrons within atoms.

Energy
The capacity for doing work and overcoming inertia.

Energy Structures
Anything consisting of matter or energy that has form.

Equidistant
Equally distant.

Equilibrium
A state of balance.

Escape Velocity
The minimum velocity an object must attain to escape a gravitational field.

Event Horizon
The boundary surrounding a black hole that separates the region of space-time from which light can escape from the region from which it cannot.

Extra-Atomic Transient Energy
Transient energy that exists outside of atoms.

Fabric of the Universe
Also known as Space-Time; that from which everything else originates.

Firefly
A nocturnal beetle that glows due to chemical reactions.

First Law of Thermodynamics
The law of conservation of energy, which states that in any interaction the total amount of energy does not change.

Fission Blast
An explosion due to a fission-type nuclear chain reaction.

Flux
Change.

Frequency
The number of times something occurs within a particular extent of time, such as an electromagnetic wave.

G
The letter that represents the gravitational constant.

G Forces
The force of acceleration due to gravity at a celestial body's surface and multiples of this force.

Galactic Scale
Relating to the size of a galaxy.

Galaxy
A vast system of stars, planets, and other matter and energy revolving around a common center of gravity.

Gamma Rays
Electromagnetic radiation at the highest-energy end of the electromagnetic spectrum.

Governing Laws
A set of laws that has jurisdiction over something and by which that something must abide.

Gravitational Bond
The unification of two or more things by way of gravity.

Gravitational Field
An extent of space in which gravity is present.

Gravitational Lensing
The distortion of the fabric of the universe by massive objects, which results in light from sources on the other side being curved around the massive object, causing the sources to appear to be at different locations than they really are.

Gravitational Potential Energy
Energy stored in an object because of its position in a gravitational field.

Gravitational Waves
Disturbances in the fabric of the universe caused by the motions of matter.

GLOSSARY

Graviton
The theoretical particle thought to be the cause of gravity.

Gravity
The force that causes objects to tend to move toward each other.

Homogeneous
Having the same composition throughout.

Inertia
A property of the universe that we experience through matter whereby objects at rest tend to remain at rest and objects in motion tend to remain in motion at the same speed and direction until accelerated by some force.

Inertial State
Relating to the condition of the inertia affecting an object.

Inevitable
Cannot be prevented from happening.

Inextricably Connected
Impossible to be disconnected.

Infinite
A mathematical abstraction that refers to a quantity that is always more than any other.

Infinite Directions
A quantity of directions without a finite limit.

Infinite Motion
Motion that is infinite in speed and direction.

Infinite Speed
A speed whereby an infinite distance can be crossed instantly.

Infinite Volume
A quantity of volume without a finite limit.

Infinitesimal
Infinitely small; without a finite limit.

Instance of Origo
Any leading edge of origo that is being taken into consideration, whether it be the inward force, outward force, or other, and whether it carries light energy or not.

Interference Patterns
The pattern formed when two or more waves interact with each other.

Interstellar Travel
Travel between stars.

Inversely Proportional
When one quantity increases, the other decreases by the same amount.

Inward Force
Origo that is moving toward anything, and in the case of an energy structure, until it reaches a plane 90 degrees from its direction of travel that intersects the center of the energy structure.

Isolated Observer
An observer that has no effect whatsoever on that which is being observed.

Isolated System
A system that is free from the influence of an external force.

Kinetic Energy
Energy that is stored in objects due to their motion.

Laterally
Sideways.

Light Year
A measurement of length equal to the distance light travels in one Earth year; about 5.9 trillion miles.

Line Spectra
Characteristic lines found in the light spectrum received from any source, often distant galaxies.

Local Speed of Light
The speed that light travels at a specific location.

Logic
The reflection of the nature of the universe.

Macroscopic
Objects large enough to be seen with the naked eye.

Magnetic Field
An extent of space in which magnetism is present.

Magnetic Monopole
A magnetic pole without an accompanying opposite pole.

Magnetic North Pole
The magnetic pole at which field lines exit the magnet.

Magnetic South Pole
The magnetic pole at which field lines enter the magnet.

Magnitude
Size.

Mass
The measure of the inertia of a body; a quantity of matter.

Matter
That which is physical and occupies space.

GLOSSARY

Medium
A surrounding or enveloping element.

Model
A representation or example of something.

Molecules
One or more atoms constituting the smallest part of an element or compound that can exist separately without losing its chemical properties.

Momentum
An objects mass times its velocity.

Morse code
A system of telegraphic symbols composed of dots and dashes or short and long flashes or beeps used to transmit messages.

Motion
The act or process of changing position.

Neutron
An electrically neutral particle of the atomic nucleus.

Neutron Star
Extremely dense stars that are the remnants of supernovae.

Nothing
The absence of anything; also, completely empty space devoid of even the fabric of the universe.

Nuclear Fission
Nuclear reactions that occur when atomic nuclei split apart.

Nuclear Fuel
Elements such as uranium-235, uranium-233, and plutonium-239 that are used to produce nuclear reactions.

Nuclear Fusion
Nuclear reactions that occur when atomic nuclei combine.

Nuclei
Plural of nucleus.

Nucleus
The central core of any system; often referring to an atomic nucleus.

Origo
The fabric of the universe, which is always in motion.

Outward Force
Origo that is moving away from anything; in the case of an energy structure, when it is moving away from a plane 90 degrees from its direction of travel that intersects the center of the energy structure.

Parameter
A fixed limit or guideline.

Particle
A small elementary component.

Permeate
To spread thoroughly through.

Photoelectric Effect
The emission of electrons from certain metals upon striking the metal with light of high enough energy.

Photoelectric Emission
The emission of electrons that takes place with the photoelectric effect.

Photons
Said to be the smallest component of light.

Physics
The study of motion, matter, and energy.

Plane
A surface such that a straight line joining any two of its points lies entirely within the surface.

Plutonium-239
Nuclear fuel commonly used in atomic bombs.

Primary System
The universe, of which all other systems are subservient.

Propagation of Light
The transmission of light.

Propulsion System
A system for propelling something.

Proton
A positively charged particle within atomic nuclei.

Quantum Entanglement
A condition wherein two separate photons are somehow linked in such a way as to allow instant communication between them.

Radially Inward
In all directions toward something.

Radially Outward
In all directions away from something.

Radiation
The emission and transmission of radiant energy, usually by radioactive substances.

Red-Shift
A phenomenon in which light is shifted toward the red end of the electromagnetic spectrum.

GLOSSARY

Red-Shift Communication System
A system that uses red-shifted light as a means of communication.

Relevant Sphere of Light
The spherical volume whose radius extends from the observer to the object being observed.

Revolution
The movement of any celestial body around a center.

Rotation
To turn on an axis.

Second Law of Motion
Newton's second law, which states the larger the mass the more force is required to accelerate it; Force = mass x acceleration.

Second Law of Thermodynamics
States that the disorder of an isolated system can never decrease.

Self-Perpetuating
Renewing oneself or itself for an indefinite length of time.

Solar Eclipse
Happens when the moon passes between the Earth and the sun, and obscures the sun from view.

Space
The volume of the universe.

Space-Time
The fabric of the universe.

Spatial Location
Something's location in space.

Spectral Shifting
In light analyzed from distant sources, the shifting of characteristic spectral lines from their normal terrestrial locations.

Speed of Origo
Equal to the speed of light.

Subatomic Particles
Small constituent particles that comprise atoms.

Superconductivity
A condition wherein electricity can flow without resistance.

Superluminal
Faster than light.

Thermonuclear Explosion
An explosion produced by a Hydrogen bomb.

Time
The measurement and comparison of the periods of separate events.

Trajectory
The path of an object in motion.

Transient Energy
Energy that is easily changed from one form to another; light, heat, etc.

Universal Constant
A definite value for some particular quantity that is the same everywhere.

Universal Phenomenon
A phenomenon that occurs everywhere.

Universe
Everything that exists, including space.

Vacillations
Swaying one way and the other.

Vacuum
A space absolutely devoid of matter.

Variations
Changes.

Velocity
Speed and direction.

Waning
Diminishing, becoming smaller.

Waxing
Increasing, becoming larger.

INDEX

A

Absolute zero, electrospheres at, 118
Acceleration, 40, 43, 54, 67
    artificial gravitational field, 49
    due to gravity, 43-46, 56
Andromeda, 93
Artificial gravitational fields, 47-51, 57
Atomic explosions, 33
Atomic model, 63, 111-124
    current model, 111, 124
    laws of physics, 117-118
    Origo and, 114-115, 122-123
    transient energy and electrosphere, 115-121, 124
Atomic nucleus. *See* Nucleus

B

Big bang theory, 94
Black hole, 101-108
    matter at speed of light, 105
    Origo and, 102, 104
Blue shift, 86, 93, 98
Bohr, Niels, 113
Bursters, 33

C

Cavendish experiment, 155-156
Closed loop theory of universe, 10-11
Compressed Origo, 129
Copernicus, Nicolaus, 146-147
Corpuscular property of light, 61, 81-84, 98
Curvature of "space-time", 41-43

D

Day, 23
Diffraction box experiment, 84-85
Doppler effect, 85, 93, 98
Doppler shifts, 85-88

E

Einstein, Albert, x, 19, 41, 63, 79, 148
Electricity, predictions about, 151-153
Electromagnetic spectrum, 53, 74
Electromagnetism, Maxwell's laws, 112
Electrons
    atomic structure and, 111, 112, 119, 124
    nature of, 121

Electrosphere, 115-118, 120, 124, 131
Empty space, 13-15, 18, 19, 22, 53, 63, 65, 68
Energy
    forms of, 71
    generation of light, 71-73
    inertia, 53-55, 57
    kinetic energy, 19
    in magnetic field, 129-131
    Origo, 18-21
    transient energy, 115-121
Energy structures, 29
Equilibrium, 54, 57
Ether, 63
Events, 4-7, 22

F

"Fabric of space-time", 41
Faraday, Michael, 147-148
Field propulsion, 55, 141, 153
First event theory of universe, 9-10
First law of thermodynamics, 72
Future (time period), 6-7

G

Galaxies, 34, 83, 87-88, 93, 96
Galilei, Galileo, 147
Gases
    electrospheres, 119
    gravitational bond between nuclei, 117
General theory of relativity, x, 41
Governing laws of the universe, 16, 22
Gravitational Challenge, 155-156

Gravitational field, 43-46
    black holes, 101-108
    light and, 90-92, 106
    of nucleus, 114-115
Gravitational lensing, 91-92
Gravitational waves, 52-53
Gravitons, 52, 57
Gravity, 39-41, 46
    acceleration due to, 43-46, 56
    artificial gravitational fields, 47-51
    Gravitational Challenge, 155-156
    Origo and, 42, 52, 56

H

Hubble, Edwin P., 85, 93, 149
Hydrogen atom, atomic structure, 111, 124

I

Inertia, 53-55, 57
Infinite motion, 20, 27
Infinite time, 13-14
Instance of Origo, 75

K

Kinetic energy, 19

L

Laws of physics, 16, 22, 34, 117-118
Light, 97-98
    age of universe, 94
    black holes, 101-108
    classical experiments with, 77-90
    corpuscular property of, 61, 81-84, 98

INDEX | 169

Light *continued*
    darkness of night sky, 95-96
    diffraction box experiment, 84-85
    Doppler shifts, 85-88, 93-94
    frequency of, 73-74, 97
    generation of, 70-73
    gravitational field and, 90-92, 106
    Michelson-Morley experiment, 79-81, 97, 148
    motion of, 74-77
    nature of, 61-62, 71
    Origo and, 69-70, 73-74, 75, 81, 82, 84, 88, 97-98, 103, 107
    packets, 82
    photons, 61, 81, 83, 84, 97, 98
    propagation of, 62-70, 74-75, 97
    red-shift communication system, 89-90
    speed limit for, 106
    speed of, 28, 62-63, 65, 70, 92, 97, 108
    wave nature of, 61, 79-81
Liquids, gravitational bond between nuclei, 117
Logic, 7-8, 16, 22

M

Magnetic field, 130
    attraction between north and south pole, 138-139
    generation of, 131-135
    repulsion between north poles, 135-137

Magnetic field *continued*
    repulsion between south poles, 137
    transient energy and, 131-135, 137, 138-139, 142
Magnetic monopoles, 140, 142
Magnets, 129-142
    attraction in, 138-139
    magnetic monopoles, 140, 142
    repulsion in, 135-137
Matter
    above the speed of light, 106-107
    composition of, 129
    destruction of, 29
    as energy structures, 29
    formation of, 28, 29-30, 34, 68-69
    in galaxies, 94
    motion of Origo through, 122-123, 124
    Origo and, 29, 34, 122-123, 124
    at speed of light, 105
Maxwell, James Clerk, 113
Maxwell's laws of electromagnetism, 112
Michelson-Morley experiment, 79-81, 97, 148
Monopoles, 140, 142
Moon, 57, 98, 108
Motion, 19
    infinite motion, 20, 27
    of light, 74-77
    Newton's second law of motion, 54
    Origo, 18-21

N

Neutron star, 123
Newton, Sir Isaac, x, 54, 143, 147
Newton's second law of motion, 54
North poles (of magnet), repulsion between, 135-137
Nuclear fission, 33
Nuclear fusion, propulsion system powered by, 49
Nucleus
   atomic structure and, 111, 117-118, 124
   gravitational field of, 114-115
   transient energy, 115

O

Origo, 18-21, 22-23, 27, 34, 44, 122
   atomic model and, 114-115, 116, 122-123
   atomic nucleus and, 114
   black hole and, 102, 104
   compressed Origo, 129
   density of, 69
   Doppler shifts, 86
   as fabric of the universe, 41-42
   gravitational field and, 43-46
   gravity and, 42, 52, 56
   instance of, 75
   inward and outward forces, 30-33, 34, 35, 43, 44
   light and, 69-70, 73-74, 75, 81, 82, 84, 88, 97-98, 103, 107
   magnetism and, 129
   matter and, 29, 34, 122-123, 124

Origo continued
   motion of through matter, 122-123, 124
   movement of, 32, 122-123, 124
   as "space-time", 41, 56
   transient energy and, 115-119

P

Packets of light, 82
Particles, light as, 61, 81-84
Past (time period), 5, 6
Photons, 61, 81, 83, 84, 97, 98
Physics, laws of, 16, 22, 27, 34, 117-118
Pound, Robert, 88
Pre-universe, 12, 18, 19, 22, 27
Present (time period), 5, 6
Propulsion systems
   with artificial gravitational fields, 47-51
   predictions about, 153

Q

Quantum entanglement, 90
Quantum theory, x, 149

R

Rebka, Glen, 88
Red shift, 86, 89, 93, 94, 98
Red-shift communication system, 89-90
Repulsion, in magnets, 135-137
Roemer, Olaus, 62

S

Second law of motion, 54
Slipher, Vesto M., 85, 149

INDEX

Solids, gravitational bond between nuclei, 117-118
South poles (of magnet), repulsion between, 137
Space, 53
   empty space, 13-15, 18, 19, 22, 53, 63, 65, 68
Space Shuttle, new propulsion system, 49, 50
"Space-time", 41-43, 56, 65
Speed of light, 28, 62-63, 65, 70, 92, 97, 108
   light-speed limit, 106
   matter above the speed of light, 106-107
   matter at, 105
Statistical mathematics, 149
Superconductivity, predictions about, 151-153
Superluminal communication system, 90, 98
Superluminal travel, 107, 108

T
Thermodynamics, first law of, 72
Time, 3-7, 22
   day, 23
   in empty space, 13-14
"Time travel", 7
Transient energy
   atomic structure and, 115-121
   magnetic field and, 131-135, 137, 138-139, 142

U
Universe, 22-23
   age of, 94
   closed loop theory of, 10-11
   empty space, 13-15, 18, 19, 22, 53, 63, 68
   first event theory of, 9-10
   governing laws, 16, 22
   matter, origin of, 28
   origin of, 9-11, 22-23, 27
   Origo, 18-21, 22-23, 94
   pre-universe, 12, 18, 19, 22, 27
   spontaneous origins of, 16-17, 22
   stability of, 29

V
Volume, 14, 22

W
Warp, of "space-time", 41-43, 56, 65
Wave nature of light, 61, 79-81

ABOUT THE AUTHOR

November 11, 1970, Michael F. Jones was born in Tacoma, Washington. With the exception of one magical year, all of his life he has resided in the beautiful state of Washington, which he loves very much.

In the autumn of 1975, his family moved to Council Bluffs, Iowa, where his father had grown up. There, Jones experienced things that would be imprinted on his memory forever: walking sticks; giant tree spiders and grasshoppers; five-foot-long, multi-colored king snakes resting in the driveway. There were storms that were nothing short of awesome. Thunder that vibrated through his body as if he were standing next to a barreling locomotive, hail the size of ping-pong balls, monsoon-like rain, lightning that turned night into day and blasted down trees in the yard, and snow three feet deep all made a lasting impression on his young mind. These wondrous things made him realize the incredible beauty of the world and the power of Mother Nature.

As a curious, young boy, he wanted to know how and why things happen. He and his older, equally adventurous brother Earl would take nearly everything apart. Getting them back together was another matter entirely!

An honor student for most of his elementary years, he excelled in science; it was his best subject. It was fun! Science opened the doors to the universe and let him in! By the time his teen years rolled around, a split family and tumultuous childhood had hardened him toward authority. He developed a very strong and independent will that led him away from the mainstream.

Even through his trying adolescent years, he never lost his great love for science and the universe. His independent studies consisted of reading books and magazines about astronomy, gravity, early scientists and their work, the Earth, the development of the industrial age, and, among other things, the manufacturing of materials and products. There was very little that did not interest him. He explored many different subjects, but astronomy, gravity, and the mysteries of the universe were the ones that pulled on his heartstrings.

Jones has been a dedicated student of the universe his entire life. He has no Ph.D. or certificate to declare to the world these many years of study. Some may think that without a certificate he should not write such a comprehensive book as *Logical Universe*. However, as a young boy, he did not allow other people's opinions to control his actions, and as a young man he continues to direct his own actions.

Jones once thought that scientists knew everything there was to know about the universe, but the more he learned, the more he realized just how little they really did know. The doors to the universe that had opened for him as a boy seemed to be closing. There was no one who could answer the questions that he wanted answered.

As a free thinker, Jones' desire to understand his universe

would not permit him to be content with modern science's feeble attempts to explain it, and his ability to understand it would not be limited by modern science's failure to understand it.

Logical Universe is an interpretation of the universe as seen through Jones' eyes.

CONTACT THE AUTHOR

Logical Universe, Questions and Comments
P.O. Box 45470
Tacoma, WA 98445

If you would like your question, comment, or story to be included on our website or in any other Inertial Press publication, the following release MUST be included with your letter:

RELEASE

I _____ (print name), do hereby certify that I am the author of the letter which accompanies this Release, and represent that the contents of said letter are my own. I authorize Inertial Press, its licensees, agents, successors and assigns, to publish in any manner of media whatsoever, the contents of my letter.

Furthermore, I consent to the use of my name, city and state in all publications by Inertial Press. If I do not wish to have my name published I will so state in writing in the comments section below. I further understand that by sending the attached letter to Inertial Press, I do not expect to receive any renumeration or compensation whatsoever from Inertial Press, its licensees, agents, successors, and assigns.

Being of lawful age, I hereby expressly release and hold harmless Inertial Press, its licensees, agents, successors, and assigns, from all liability, including, but not limited to, any and all claims for damages for libel, slander, defamation, invasion of privacy, or any other claim based on the use of the above referenced letter.

_____

(your signature)

Address: _____

_____

Phone number: _____

Comments: _____

_____

_____

ORDER FORM

Fax orders: 253-531-3010; Send this form.

Postal orders: Inertial Press, P.O. Box 45470
Tacoma, WA 98445, USA. Send this form.

Telephone: 888-531-2010

Prints of all *Logical Universe* illustrations are available at:
www.inertialpress.com

Name: _____

Address: _____

City: _____ State: _____ Zip: _____

Telephone: _____

Email address: _____

Please Send the Following Copies of Logical Universe:

☐ I would like to receive notice of future Inertial Press offerings.

Quantity: _____

Quantity x $22.95 = _____

(Please add 8.8% for products shipped to Washington addresses.) Sales tax: _____

Shipping: _____

SHIPPING BY AIR
U.S.: *$4.50 for first book and $2.50 for each additional book.*
International: *$9.50 for first book and $5.00 for each additional book (estimate).*

TOTAL: _____

PAYMENT: ☐ Check ☐ Credit card:

☐ Visa ☐ MasterCard ☐ AMEX ☐ Discover ☐ Other _____

Card number: _____

Name on card: _____ Exp. Date: _____

Signature _____